T0170196

A Framework for Precision Conventional Strike in Post–Cold War Military Strategy

John Birkler
Myron Hura
David Shlapak
David Frelinger
Gary McLeod
Glenn Kent
John Matsumura
James Chiesa
Bruce Davis

NATIONAL DEFENSE RESEARCH INSTITUTE

RAND

Prepared for the Commission on Roles & Missions of
The Armed Forces

Precision conventional strike (PCS) is the practice of attacking selected targets with sufficient accuracy for high probability of kill
and low collateral damage. The purpose of this report is to offer
decisionmakers and analysts a framework for making first-order
assessments regarding priorities for the development and acquisition
of precision-conventional-strike (PCS) weapons. Specifically, the
report

- identifies key objectives to which PCS weapons may contribute

- assesses the applicability of currently available and programmed
  PCS weapons across four scenarios

- suggests priorities for future acquisition and development of PCS
  weapons.

The report also suggests ways of promoting innovation in the development of new PCS weapons and concepts. However, we do not
pretend to resolve those issues, nor do we recommend a plan of
action for PCS system development and acquisition.

This report originated in one of several tasks related to deep attack—
attack well beyond the proximity of friendly troops—and PCS that
were undertaken at the request of the Commission on Roles and
Missions of the Armed Forces. RAND was asked to address the issue
of strategy. Other tasks (addressing weapon systems, organization,
and so on) were performed by the Institute for Defense Analyses and
the Center for Naval Analyses. This report is intended for people in

the Office of the Secretary of Defense, the Services, and industry interested in PCS issues.

PCS was one of several areas in which RAND provided analytic support to the Commission's deliberations. The Commission was created in 1993 by Congress to review and evaluate "current allocations among the Armed Forces of roles, missions, and functions" and to "make recommendations for changes in the current definition and distribution of those roles, missions, and functions" (National Defense Authorization Act for FY 1994, *Conference Report,* p. 198).

The Commission does not necessarily endorse the options presented, the methodology involved, or the discussion contained in this report. This represents one of many inputs provided to inform the deliberations of the commissioners, who applied their own experience and judgment in arriving at the conclusions and recommendations that are found in the Commission's final report, *Directions for Defense.*

Analytic support to the Commission was conducted within the National Defense Research Institute, a federally funded research and development center sponsored by the Office of the Secretary of Defense, the Joint Staff, and the defense agencies.

CONTENTS

Appendix

FIGURE

TABLES

Precision conventional strike (PCS) is the practice of attacking selected targets with sufficient accuracy for high probability of kill and low collateral damage. Today's precision-conventional-strike weapons were developed for the primary purpose of fighting a major war against the Soviet Union. *What value do they have in future military strategies?* The answer to this question will help to shape the roles of these weapons in future U.S. military campaigns and will have a bearing on whether some campaigns may even be undertaken. In this report, we present a methodology to assist in arriving at an answer and draw some preliminary lessons from illustrative applications of that approach.

KEY FINDINGS

We found that:

- Existing weapons provide fairly robust capabilities against soft and semihardened fixed structures, stationary mobile targets, and some targets moving with predictable direction and speed.

- However, the effectiveness of existing weapons may be limited by weather, by availability of intelligence on targets and on routes to targets, and by enemy countermeasures such as navigation signal jamming.

- Furthermore, where terminal air defenses have not been suppressed and air superiority has not been established, existing weapons cannot be effectively delivered against hardened targets and armor unless stealth aircraft are employed.

- As a result of these limitations, PCS weapons today cannot always make major contributions to achieving campaign objectives as diverse as suppressing war-supporting infrastructure and halting invading armies.

We emphasize the preliminary, illustrative nature of our exercise and its reliance on previous studies. Clearly, much can be derived from more-comprehensive, systematic implementation of the methodology, supported by new analyses of weapon and delivery platform capabilities, including intelligence support and other infrastructural elements. Nonetheless, we drew from our first-cut analyses of PCS capabilities and shortcomings some potentially useful inferences regarding possible avenues for the investment of system development and acquisition dollars:

- Over the near term, system development dollars should be directed toward alleviating the limitations of weather, intelligence support, and jamming.

- Progress on new antitank weapons should be carefully monitored. If these weapons perform as advertised, they could contribute mightily to the campaign objective of halting advancing armies, and sufficient numbers should be procured as a matter of high priority.

We did not attempt within the limited scope of our study to rank the benefits or assess the costs of the potential solution directions we considered. These, too, are important issues for further research. Let us now elaborate on the framework and its ramifications.

CONTEXT, EFFECTIVENESS, SHORTCOMINGS

Our approach to assessing the value of PCS in future military strategy accounts for the fact that various scenario-related factors—collateral-damage constraints, weather, enemy action, intelligence preparation—influence the appropriateness of PCS weapons for targets designated for destruction. These factors can also interact with each other to make PCS infeasible. For urban targets, collateral-damage constraints may weigh against the use of all but the most-accurate weapons, e.g., laser-guided bombs. But bad weather can prevent the use of such weapons. (Of course, bad weather can also inhibit platform operation.) A robust enemy air defense system can have a similar effect on nonstealthy delivery aircraft, since laser-guided bombs (and certain other PCS weapons) must be released in the vicinity of the target. The enemy can also decrease the effectiveness of GPS-guided standoff weapons by jamming the GPS signal frequencies. Finally, the applicability of PCS systems also depends on the availability of intelligence about the target. All PCS weapons require accurate data on target location and other characteristics; autonomous weapons may also require special target imagery and information about the path to the target.

Although PCS effectiveness depends on the weapon, context, target, and scenario, some generalizations are possible. Existing weapons provide fairly robust capabilities against soft and semihardened fixed structures, stationary mobile targets, and some targets moving with predictable direction and speed. By providing the capability to conduct precise, effective strikes against critical targets, these weapons can increase the effectiveness of military operations. For example, air-delivered PCS weapons can reduce the exposure of flight crews to enemy defenses by permitting higher delivery profiles, by allowing greater standoff, and by reducing the number of sorties required. Exposure reduction and sortie reduction will result in cost savings, as will the need to transport a lesser weight of ordnance into or near the theater. Such savings should be deducted from the well-known high unit cost of PCS weapons when assessing their true costliness. But the issue of costliness raises another question with ramifications for all the preceding benefits: Are current stocks of PCS systems of various types sufficient to achieve campaign objectives within reasonable time constraints? We do not address force structure in this report, except in very general terms when considering long-term force evolution. Weapon stocks thus represent another topic that would have to be part of a more comprehensive investigation of the value of PCS in future strategy.

What are the shortcomings of current and planned PCS systems? Contextual limitations—weather, enemy countermeasures—have already been mentioned. Aside from those, PCS systems lack the capability to destroy a substantial number of very hard targets and deeply buried targets. Air-launched PCS weapons with sufficient standoff range to avoid terminal defenses are ineffective against armor, so those defenses must be suppressed before tanks and other armored vehicles can be attacked effectively with low risk. Soft targets moving with uncertain direction and speed are also problematic, because of the difficulty of identifying and tracking such vehicles long enough to hit them with a standoff weapon. As mentioned above, even fixed-target attack places a substantial burden on intelligence collection and analysis assets if autonomous weapons such as cruise missiles are to be used. Furthermore, the high cost of many PCS systems restricts the numbers that can be acquired and focuses their use on a limited number of targets of high individual value.

These generalizations about targets can in turn be translated into first-order lessons about the contributions of PCS systems to various tasks needed to achieve operational objectives. We addressed a broad range of such tasks, for example, attacking enemy air defenses, disrupting electric-power production, providing long-range supporting fires. As might be expected from the preceding paragraphs, this exercise suggests that care be taken not to overestimate the leverage gained from PCS. For example, while the United States achieved its objective of suppressing the enemy's war-supporting infrastructure in Operation Desert Storm, achieving the same objective where weather is a more serious limiting factor could be problematic. Halting an invading army with PCS weapons could also be difficult if mobile air defense units moved with tank columns, since current standoff weapons are not very effective against moving armor.

IMPLICATIONS FOR SYSTEM DEVELOPMENT AND ACQUISITION

Can system development dollars be directed to overcome such limitations? To some extent, this is already being done. DoD is making a significant investment in the development of PCS weapon/submunition combinations intended to kill tanks, e.g., Sensor-Fuzed Weapons (SFWs) using Wind-Corrected Munition Dispensers (WCMDs), the Joint Standoff Weapon (JSOW) with SFW submunitions, and the Army Tactical Missile System with Brilliant Antiarmor Submunitions. It will be important for decisionmakers to closely monitor these developments to ensure that these systems work as advertised. Meanwhile, DoD already has a system—the GBU-28—that is effective against hard targets and buried targets. However, there are very few in the inventory and air defenses must be suppressed for low-risk delivery, because the only aircraft that can deliver the GBU-28 are not stealthy. If such targets are to be attacked, procurement of additional GBU-28s is a near-term option; for the farther term, options include developing a standoff weapon. Additional priorities include the development of an all-weather, high-accuracy PCS weapon; reduction of vulnerability to GPS jamming; and the development of a more-effective intelligence support infrastructure.

But even if those issues can be addressed, the utility of PCS systems will still be limited by their high unit cost, particularly if current budgetary limitations persist.[1] If technological advances permit the development of weapons that can be produced inexpensively and delivered very accurately (the Joint Direct Attack Munition, JSOW, and WCMD are steps in this direction), different, expanded roles for PCS systems might emerge. Such roles could be facilitated by the evolution of new concepts of operations, e.g., in-flight reprogramming of stealth platforms and separation of PCS hunter and killer functions.

Those last two possibilities in particular raise once again the question of Service roles and functions. Who will "own" the PCS forces of the future? We opt for a joint perspective. We propose a new framework to make informed choices among promising new weapon concepts—choices based on the merits of cases, unhampered by preconceived notions of which roles and functions are "assigned" to a particular Service. Specifically, we propose dissolving the current Service-oriented marriage of platform, weapon, and munition and allowing, for example, the Air Force to supply a munition that might equip a Navy weapon adapted for launch from an Army platform.

CONCLUSION

DoD is supporting a number of development efforts that might generate large payoffs in the future. However, any commitment of large

---

[1]In referring to the consequences of "high unit costs," we are observing that large numbers of weapons with high budgetary costs are unlikely to be bought. As mentioned above, true cost comparisons should balance the value of lower attrition, a reduced logistics tail, and other factors not normally counted in procurement costs. Also, high-cost systems may be more economical than lower-cost systems if their benefits are proportionately higher or unachievable by less-expensive systems.

amounts of resources to PCS systems should be accompanied by a wide-ranging, thorough analysis of the potential costs and benefits of various PCS alternatives—and alternatives to PCS. Such analysis should be carried out in light of possible long-term trends in the use of U.S. forces and in the reaction of U.S. adversaries to the persistent U.S. search for technological advantage—and in recognition of the political goals that U.S. forces are meant to achieve. On these last two points, for example, it is important to recognize the ways in which evolution of enemy countermeasures over the long term might be able to limit the potential of PCS systems. It is also important to recognize that certain objectives, e.g., changing the behavior of high-level enemy political and military leaders, cannot reliably be achieved with any conventional weapon, regardless of how accurate.

ACKNOWLEDGMENTS

This report owes much to the guidance of Col David Deptula, USAF, who framed the overarching issues to which it responds. We also acknowledge the assistance of the four Service representatives who worked with us: CDR Kevin Baxter, USN; Maj. Mark Bolin, USMC; Maj Jeff Latas, USAF; and LTC Richard Whitaker, USA. While these officers may not agree completely with either the tenets or conclusions of this report, their experience, knowledge, and cooperation were invaluable in its preparation.

At RAND, we benefited greatly from the direction and support given by David Chu, director of the Washington Research Department. We are also indebted to David Ochmanek and Leland Joe for their constructive comments and suggestions on earlier drafts. Their insights, suggestions, and encouragement were essential to completing this research. Of course, we alone are responsible for any errors of omission or commission.

ACRONYMS AND ABBREVIATIONS

AAA	Anit-aircraft artillery
AGM	Air-to-ground munition
AIL	Airborne Instruments Laboratories (a contractor)
APAM	Antipersonnel Antimaterial [weapon]
ATACMS	Army Tactical Missile System
ATR	Automatic target recognition
AWACS	Airborne Warning and Control System
BAT	Brilliant Antiarmor Submunition (formerly Brilliant Anti-Tank [submunition])
BDA	Battle damage assessment
BLU	Bomb, live unit
CALCM	Conventional Air-Launched Cruise Missile
CBU	Cluster bomb unit
CEB	Combined-Effects Bomblet
CEM	Combined-Effects Munition
CEP	Circular error probable
C3I	Command, control, communications, and intelligence
CINC	Commander in chief
CMSA	Cruise Missile Support Activity
DAB	Defense Advisory Board
DMA	Defense Mapping Agency
DSMAC	Digital Scene-Matching Area Correlator
EO	Electro-optical
FEBA	Forward edge of the battle area
GBU	Guided bomb unit
GNC	Guidance, navigation, and control
GPS	Global Positioning System

HARM	High-Speed Antiradiation Missile
HPM	High-power microwave
HUMINT	Human-source intelligence
ICBM	Intercontinental ballistic missile
IFF	Identification, friend or foe
IGPS	Integrated Global Positioning System
INS	Inertial navigation system
IR	Infrared
IW	Information warfare
JDAM	Joint Direct Attack Munition
JASSM	Joint Air-to-Surface Standoff Missile
JOG	Joint Operations Graphics
JROC	Joint Requirements Oversight Council
JSOW	Joint Standoff Weapon
JSTARS	Joint Surveillance [and] Target Attack Radar System
LANTIRN	Low-Altitude Navigation and Targeting System for Night
LGB	Laser-guided bomb
LOE	Level of effort
LRC	Lesser regional contingency
MITL	Man in the loop
MLRS	Multiple-Launch Rocket System
MNS	Mission need statement
MRC	Major regional contingency
ODS	Operation Desert Storm
OMG	Operational maneuver group
ONC	Operational Navigation Chart
OOTW	Operation other than war
P3I	Preplanned product improvement
PCS	Precision conventional strike
PGW	Precision-guided weapon
PIP	Product Improvement Program
POL	Petroleum, oil, and lubricants
PPDB	Point Positioning Data Base
SAR	Synthetic aperture radar
SAM	Surface-to-air missile
SFW	Sensor-Fuzed Weapon
SIGINT	Signals intelligence
SKEET	"Smart" anti-armor submunition
SLAM	Standoff Land Attack Missile

SLAM-ER	Standoff Land Attack Missile–Expanded Response
SLBM	Submarine-launched ballistic missile
SOF	Special-operations forces
TEL	Transporter-erector-launcher
TERCOM	Terrain Contour Matching
TLAM	Tomahawk Land Attack Missile
TLE	Target location error
TMD	Tactical Munitions Dispenser
TSSAM	Tri-Service Standoff Attack Missile
UAV	Unmanned aerial vehicle
USA	United States Army
USAF	United States Air Force
USMC	United States Marine Corps
USN	United States Navy
WCMD	Wind-Corrected Munitions Dispenser
WGS	World Geodetic System

Chapter One

INTRODUCTION

The purpose of this report is to offer decisionmakers and analysts a framework for making first-order assessments regarding priorities for the development and acquisition of precision-conventional-strike (PCS) weapons. Specifically, the report

- identifies key objectives to which PCS weapons may contribute
- assesses the applicability of currently available and programmed PCS weapons across four scenarios
- suggests priorities for future acquisition and development of PCS weapons.

The report also suggests ways of promoting innovation in the development of new PCS weapons and concepts. However, we do not pretend to resolve those issues, nor do we recommend a plan of action for PCS system development and acquisition.

Precision conventional strike (PCS) is the practice of attacking selected targets with high accuracy and limited collateral damage. It is achieved by a variety of weapon systems collectively known as precision-guided weapons (PGWs). PCS can encompass targets at the tactical, operational, and strategic levels of war and can be conducted by force elements fielded by all four armed services. (While it has been typically associated with deep attack, that association is not a necessary one.)

Like living organisms, current PCS weapon systems were shaped by their environment and selected through a process that accentuated certain qualities—qualities that are immutable over the short term.

1

Many of the strengths and limitations of today's PGWs came into being quite early in the design process, when the weapon designer's concepts were given form. PCS systems were developed primarily to counter a modern industrialized adversary, such as the Soviet Union, deploying armor and other mechanized forces. That type of adversary would be rich in high-value targets suitable for PCS attack.

A major issue is the role of PCS against a different type of adversary, say, one based on infantry or an agricultural economy where there are many potential targets, none of which is particularly critical or valuable. As the world and potential conflict sites shift, we need to ask what strategies are appropriate to the use of these systems in the post–Cold War environment? Do we need a set of weapons with an area capability, and will we need to devise a strategy to avoid massive collateral damage? What are the implications of these strategies for the resource decisions involved in future development and acquisition of PCS systems? How do our current systems fare in this new environment? These are the questions we undertake to address in this report. We do not attempt to provide more than first-order answers; instead, we lay out some of the factors that need to be taken into account and some options that bear consideration.

ROLES, MISSIONS, AND FUNCTIONS

The Commission on Roles and Missions of the Armed Forces was directed by law to review the allocation of "roles, missions, and functions" among U.S. forces. Technically, "functions" refer to such activities as recruiting, supplying, training, mobilizing, and demobilizing the armed forces and are carried out by the military departments.[1] "Roles" are the parts played by particular DoD elements; most obviously, a role may be to carry out a particular function. For example, it may be the role of one of the armed services to acquire and field a particular PGW. "Missions" refer to sets of operations assigned to the combatant commanders, e.g., defense of the U.S. homeland, warfare against other nations, peacekeeping.

---

[1]These definitions are based on PL 99-483 (the Goldwater-Nichols Act of 1986), Sec. 3013.

This report is concerned with strategy. It is informed by RAND's strategy-to-tasks framework, a hierarchical view of decisionmaking and execution in which strategies at one level become objectives for tasks to be carried out at the next level down.[2] Thus, national goals such as providing for the common defense are fulfilled by strategies such as defeating aggression against the United States and its allies. The latter is also viewed as a national security objective, achieved through more specifically stated strategies, e.g., defeating large-scale aggression by North Korea against South Korea. The latter strategies form national military objectives to be fulfilled through campaign strategies such as gaining air superiority, and so on down to such specific operational tasks as damaging key hardened air base support facilities.

Missions, as defined above, fall within this framework. They embody objectives (or strategies) that are among the links in the logical chain through which national goals drive operational tasks. Thus, when we speak of strategy, we speak of missions, tasks undertaken to achieve missions, or objectives that motivate them. Roles and functions are not as closely related to strategy, so we do not speak much of them. Certainly, if a strategy is to be carried out, someone must provide the wherewithal in terms of troops and equipment, along with the doctrine for employing them. But while such functions should be dictated by strategy, they are not themselves strategy, and they should not (over the time frame considered here) motivate strategy.

DIFFERING PERSPECTIVES ON DEEP ATTACK

There are currently differing, polarized views on how precision strike should be organized in the future. One view essentially calls for a "partitioning" of the battlefield into deep, close, and rear areas. Services could then choose areas of responsibility relevant to what they view as their roles. Such a partition would reduce cross-Service coordination difficulties and airspace deconfliction challenges, resulting in a much more fluid and efficient deep attack campaign. The other view suggests that the future battlefield is perhaps becoming increasingly "dynamic" and nonlinear in nature, resulting in in-

---

[2]See D. E. Thaler, *Strategies-to-Tasks: A Framework Linking Means and Ends*, MR-300-AF, RAND, Santa Monica, Calif., 1993.

distinct deep, close, and rear areas. In this case, a "partitioning" approach would not be viable; instead, increased efforts should be made to ensure more-successful joint operation across the span and depth of the battlefield. New technologies promoting digitization and deconfliction (for simultaneity) could be "enablers." This disparity in views still remains analytically unresolved.

Although we considered addressing this important issue, limitations on both scope and time of this study precluded any substantive analysis. For the purposes of this report, we lean toward the joint view. However, we feel that analytic work can and should be applied in the future to help resolve this issue.

SCOPE OF THIS REPORT

The scope of this report derives from the evaluation plan for deep attack and PCS that was issued by the staff of the Commission on Roles and Missions of the Armed Forces. That plan assigned questions of organization, systems, and force structure to other team members. RAND alone was to focus on "assessment of PCS weapon systems in terms of their potential role in a future military strategy." *It is not our purpose here to draw conclusions about the assignment of roles and missions, the effectiveness of PCS weapons, or the most cost-effective PCS force structure.* We take up organizational and system issues only insofar as they are related to the central research questions cited above.

In Chapter Two, we break strategies into tasks and discuss some operational issues surrounding PCS use. We outline key scenario variables and intelligence requirements and how they affect the role that PCS systems can play in future campaigns. In Chapter Three, we discuss the relation between the strategic future for PGWs and options for system development. Finally, in Chapter Four, we suggest a plan for acquiring PCS systems that is consistent with the strategy-driven framework established in earlier chapters. We also include three appendices. The full list of tasks we considered for precision conventional strike is given in Appendix A. In Appendix B, we elaborate on the information given in Chapter Two on intelligence support and mission-planning needs for PGWs. In Appendix C, we fully discuss four topics mentioned briefly in Chapter Three.

THE ROLE OF PRECISION STRIKE IN FUTURE CAMPAIGN STRATEGY

We began our assessment of the role of PCS systems by defining a broad set of campaign objectives, e.g., gaining and maintaining air superiority, gaining and maintaining sea control or denial, affecting the will of the enemy leadership and forces. For each of these objectives, we identified operational objectives and tasks to achieve them. For example, operational objectives supporting a campaign objective of halting invading armor are to delay, damage, or disrupt lead elements of invading armies and to delay or damage enemy forces and logistics in the rear. The lead-element disruption objective is achieved through such tasks as destroying armored vehicles on the attack, mining key routes of advance, and suppressing forward-area rearm-and-refuel points for attack helicopters. (The full set of objectives and tasks is given in Appendix A.)

Separately, we identified key operational and environmental variables that influence PCS effectiveness against a variety of target types, along with important aspects of intelligence support and mission planning. We also set out various conflict scenarios in which the effects of these factors might differ. This exercise allowed us to identify some tasks that were particularly interesting and assess the potential of precision strike in accomplishing these tasks.

We did not attempt to rank tasks because priorities assigned to tasks are scenario-dependent. For example, in the canonical Southwest Asia scenario, a high planning and operational priority is attached to the campaign objective of halting invading armies and its supporting tasks—destroying moving armor, damaging bridges, etc. On the other hand, if Iraq's strategy for gaining control of the oil fields includes use of ballistic missiles, the primary task would be to destroy surface-to-surface missile launchers.

We begin this chapter by defining the major PGW categories we considered for PCS. Then, we summarize our analysis of critical variables and proceed to the assessment of U.S. capabilities to perform key tasks.

PGW CATEGORIES

In gauging their effectiveness in different scenarios in broad terms, PGWs can be grouped into six categories (see Table 2.1).[1]

Table 2.1

Precision-Guided-Weapon Categories[a]

Category	Examples
1. Man-in-the-loop	
Laser-guided	GBUs-10, 12, 16; GBUs-24, 27; GBU-28, Hellfire
Terminal sensor	Maverick, GBU-15, AGM-130, SLAM, HAVE NAP
2. GPS-aided INS alone	CALCM, JDAM,[b] ATACMS,[c,d] JSOW,[b,c] WCMD[b,c]
3. TERCOM or GPS-aided INS plus scene-matching sensor	TLAM-C Blk II and III, TLAM-D[c] Blk II and III
4. GPS-aided INS plus target-imaging sensor	TLAM-C Blk IV,[b] JDAM PIP,[e,f] JSOW P3I,[e] JSSM,[e]
5. Anti-emitters	HARM
6. Smart submunitions	SFW carried by TMD, WCMD,[b] or JSOW; BAT[b] carried by ATACMS

[a]See p. xxi for definition of abbreviations; see text for weapon system descriptions.

[b]Under development.

[c]Dispenses submunitions (those not so noted in categories 1 through 5 are unitary weapons).

[d]INS only (future extended-range variants will have GPS).

[e]Proposed.

[f]Formerly known as JDAM III.

---

[1]Antiship missiles could constitute a seventh category, but because this report focuses on land combat, we do not include them in our analysis.

The first category consists of man-in-the-loop weapons, which are of two types:

- Laser-guided bombs (LGBs) rely on an aircrew member or ground spotter using a laser designator to illuminate the target, the reflected signal from which guides the weapon (e.g., GBU-27[2]).

- Other man-in-the-loop weapons have an onboard forward-looking sensor and are linked for data transmittal to and from the launch platform, enabling an aircrew member to guide the weapon to the target. Example:

  – The Standoff Land Attack Missile (SLAM) is a Harpoon anti-ship missile modified for use against land targets. It employs a man-in-the-loop infrared guidance system and can be used against soft fixed targets and buildings (such as power-plant generator halls). SLAM-ER (expanded response) will be a longer-range variant.

PGWs in categories 2 through 6 are autonomous weapons; once the weapon is released from the delivery aircraft (or ground launcher), it guides itself to the target without further operator assistance. This allows the delivery platform to reduce its exposure time to any threats in the target area. If the weapon has substantial standoff range (such as long-range cruise missiles have), the delivery platform can avoid not only the threats near the target but also those en route.

- Category 2: Autonomous weapons relying only on an inertial navigation system (INS), updated by the Global Positioning System (GPS),[3] for guidance to the target. Examples:

---

[2]Guided bomb units (GBUs) are a family of air-to-surface weapons with some form of guidance to enhance their accuracy. Some GBUs are laser-guided; others employ television or electro-optical (EO) guidance.

[3]An inertial navigation system (INS) is one that uses an accelerometer and gyroscope to determine position by keeping track of distance and direction traveled from a known point of origin. A GPS guidance system determines position in three dimensions by calculating differences in the travel time of signals emitted at known times from known positions by satellites in the Global Positioning System. INSs are subject to accumulated drift error; the more-accurate GPS guidance system permits error correction (but is too susceptible to jamming for use alone). For additional

- The Army Tactical Missile System (ATACMS) is a medium-range surface-to-surface missile. Fired from the Multiple-Launch Rocket System (MLRS), the currently fielded Block I system is armed with several hundred Antipersonnel Antimaterial (APAM) cluster bomblets and is used against soft area targets. The Block IA missiles will carry a smaller warhead to about twice the range (in excess of 200 km). ATACMS equipped with Brilliant Antiarmor Submunitions (BATs) are category 6 weapons.

- The Conventional Air-Launched Cruise Missile (CALCM) is a long-range standoff weapon launched from B-52 bombers. CALCM is a conversion of a nuclear-armed ALCM and carries a high-explosive warhead effective against soft point targets. Range is in excess of 1,000 miles.

- The Joint Direct Attack Munition (JDAM) is being developed to provide the Air Force and Navy with an accurate, fire-and-forget, all-weather bomb. JDAM I will consist of an inertial navigation/GPS guidance kit attached to a (previously produced) 1000- or 2000-pound general-purpose bomb and will be fielded around 1999. JDAM PIP (formerly JDAM III) will incorporate a precision all-weather seeker or other device conferring increased accuracy; a category 4 weapon (see below), it is expected to come on-line after the year 2000.

- The Joint Standoff Weapon (JSOW) is an Air Force–Navy project to deploy an unpowered glide bomb with a submunition warhead. The Navy variant will dispense a large number of Combined-Effects Bomblets (CEBs). It is scheduled to become operational around the turn of the century and will be employed against soft area and relocatable targets. The Air Force variant will be equipped with Sensor-Fuzed Weapon (SFW) submunitions and will be a category 6 weapon. A product-improved JSOW (JSOW P3I) with a unitary warhead, terminal sensor, and data-link control will be a category 4 weapon.

---

information about GPS, see Scott Pace et al., *The Global Positioning System: Assessing National Policies*, MR-614-OSTP, RAND, Santa Monica, Calif., 1995.

- The Wind-Corrected Munitions Dispenser (WCMD) is a GPS-aided INS upgrade to the unguided Tactical Munitions Dispenser (TMD). TMDs carry and dispense a variety of submunitions such as CEBs and Gator mines.[4] The addition of GPS-aided INS will improve the weapon's effectiveness (kills per pass), adverse-weather capability, and accuracy (mid-course wind correction), especially when released from high altitudes (e.g., greater than 20 kft). TMDs and WCMDs with SFW submunitions are category 6 weapons.

- Category 3: Autonomous weapons relying on terrain-aided INS or GPS-aided INS for en route navigation, along with a downward-looking sensor and a scene-matching algorithm for terminal guidance to the target. There is only one weapon in this category:

  - The Tomahawk Land Attack Missile (TLAM) is a ship- or submarine-launched long-range conventional cruise missile. TLAM-C is employed against point targets such as structures and buildings; it carries a 1000-pound warhead and has a range in excess of 500 miles. TLAM-D carries 166 CEBs to about the same range and is used against soft area targets. The Block II TLAMs use the Terrain Contour Matching (TERCOM) system (TERCOM maps, algorithm, and radar altimeter) for updating the INS during en route navigation. The Block III variants, now in production, use GPS for updating and have longer ranges. The scene-matching algorithm used by TLAM-C for terminal guidance is known as the Digital Scene-Matching Area Correlator (DSMAC). The next version of TLAM, called Block IV, will have increased accuracy and will be a category 4 weapon.

- Category 4: Autonomous weapons relying on GPS-aided INS for en route navigation and an onboard target-imaging sensor and template-matching algorithm to acquire and home on the target.

---

[4]The following should make clearer the various names associated with TMDs. A Combined-Effects Munition (CEM) (i.e., CBU-87) is a TMD that carries 202 CEBs (i.e., BLU-97 submunitions). A Sensor-Fuzed Weapon (CBU-97) is a TMD that carries 10 SFW submunitions (i.e., BLU-108 submunitions). Each BLU-108 submunition carries 4 SKEET anti-armor projectiles (the SKEET also contains the infrared sensor that detects and acquires the target).

There are no operational weapons in this category; examples of planned systems include the following:

- The proposed JDAM Product Improvement Program (PIP) variant (see JDAM, above). (Note that options provided for improved accuracy that do not employ an imaging sensor are also being considered.)

- The proposed JSOW Preplanned Product Improvement (P3I) variant (see JSOW, above).

- TLAM-C Block IV, now under development (see TLAM, above).

- The Tri-Service Standoff Attack Missile (TSSAM) was a joint effort to produce a stealthy, medium-range (more than 100 miles) missile for carriage by a variety of platforms, including B-52 and B-1 bombers. It was to carry a unitary warhead. TSSAM was canceled by the Secretary of Defense in late 1994. Its planned replacement is the Joint Air-to-Surface Standoff Missile (JASSM); that weapon will not be available until after 2000.

- Category 5: Anti-emitter PGWs rely on an onboard receiver, signal processor, and associated algorithm to detect, identify, acquire, and home on emitting targets, such as tracking radars. Example:

  - The High-Speed Antiradiation Missile (HARM) is an air-to-surface missile employed by USAF, USN, and USMC aircraft against enemy air-defense radars. It carries a fragmentation warhead, has a range of some 30 miles, and homes on radar emissions.

- Category 6: We define a sixth PGW category for weapons that derive their precision primarily from the smart or brilliant submunitions they carry.

  - SFW submunitions (carrying SKEET projectiles, which use infrared sensors to detect and acquire their targets) are relatively "smart." Ten SFW submunitions (BLU-108s), each dispensing four SKEETs, are carried by TMDs. They will also be carried by WCMDs and JSOWs.

–   BAT, which uses acoustic and infrared sensors to detect, ac-
    quire, and home on its target, will be a "brilliant" submuni-
    tion. ATACMS Block II will be equipped with 13 BATs, which
    will enable the weapon to destroy moving armored targets at
    ranges similar to that reached by the Block I system. Block
    IIA will carry six preplanned-product-improved BATs (BAT
    P3I)—able to strike "cold" stationary targets—to a range
    comparable to that of the Block IA missile.

VARIABLES INFLUENCING PCS EFFECTIVENESS

PCS weapon systems were designed for specific uses over specific
ranges of environmental variables. It is not surprising, therefore, that
their effectiveness should be very context-dependent. Table 2.2 lists
factors impinging on PCS effectiveness and shows, in a very rough
way, how they can vary with scenario. To examine factor variability,
we looked at three scenarios—a Persian Gulf scenario that is generally favorable to employment of PGWs; a somewhat more challenging—in terms of weather, topography, and targets—Korean scenario;
and a Balkans scenario to place stress on PCS assets in terms of
sensitivity to collateral damage and ambiguous targeting problems.
(Of course, there can be, for example, different Balkans scenarios
with different parameter values. The values shown are intended to
be more-or-less representative.)

It is worth noting that the variables potentially impinging on PCS effectiveness span a wide range. Some reflect technical limitations of a
system such as not being able to function in bad weather because of
the absorption and scattering of light. Others represent exogenous
factors such as the political environment, which may affect the suit-
ability of a weapon system for a particular circumstance. Of the variables listed, we identified five that strongly affect the performance
and thus the usability of many PCS systems:

• **Collateral-damage tolerance.** The tolerance of political and
  military authorities toward collateral damage can influence PCS
  effectiveness in several ways. In an effort to minimize damage to
  unintended targets, the choice of PCS systems may be restricted
  to the most accurate ones—those having a very low probability of
  impacting away from their aim point. Or, selected PCS systems

Table 2.2

Scenario Variables Influencing PCS Effectiveness
(1995 Time Frame)

	Theater		
Variable	Iraq	Korea	Balkans
Critical Variables			
Collateral-damage tolerance	Moderate	Moderate	Low
Weather	Good	Poor	Medium
Enemy air defense capabilities	Medium	Medium	Low
Enemy countermeasures/tactics	Medium	High	Low
Level of intelligence	Varies	Varies	Varies
Other Important Variables			
Level of conflict	MRC	MRC	LRC/OOTW
Prepositioning	Semi-expedi-tionary	Prepositioned	Semi-expeditionary
Infrastructure			
Air base physical capability	Good	Good	Good
Air base access	Good	Good	Medium
Naval access	Bad	Good	Good
Air base vulnerability to TBMs	Medium	High	Low
Topography	Open/Flat	Hilly/Forested	Hilly/Urban/Forested
Coalition capabilities/robustness	Low	High	Medium
War aims (U.S.)	Defeat	Defeat	Peace-keeping
Territorial sensitivity (trade land for time)	Moderate	Low	High
Enemy dependence on sophisticated systems	Medium	Medium/Low	Low
Nonlinearity of battlefield	Friendly OMGs operating	Pockets of enemy SOF	No defined FEBA

NOTES: The purpose of this table is to establish some different scenarios that are more-or-less consistent with possible contingencies the United States could face in key trouble spots. For a given column, values of the factors could, of course, differ between specific manifestations of the contingency cited and (over time) within them; also, they interact, and some, obviously, are under the opponent's control. (Level of intelligence, however, can vary so much with target class, resources expended, etc., that we do not even specify a nominal value.)

See p. xxi for definition of abbreviations.

might be limited to those with smaller warheads to limit the chance of blast and other effects damaging nearby facilities. Either way, constrained choices could hold down the number of targets attacked or the rate at which U.S. forces can attack them. The selected weapons might also be employed in accordance with very strict rules of engagement that would limit the chance of unintended damage in two ways: ensuring that the correct target is designated for attack and that the weapon will abort its mission if a problem with its guidance system occurs. The former would restrict the situations in which PCS systems can be used and the latter, the weapons that can be used.

- **Weather** can have a profound influence on the effectiveness of some PCS systems by degrading the ability of target-imaging sensors or platform-based sensors to see the target. It is also the only factor that cannot be influenced in any way by the combatants. Poor weather can render unusable some infrared-guided weapons and those relying on electro-optical sensors, including LGBs.

- **Enemy air defenses** affect the utility of PCS systems by threatening the delivery platform, the weapon, or both. Defenses might be sufficiently heavy to rule out specific platform-munition combinations or to require suppression ahead of time. Or, they could influence the tactics necessary for safe delivery of the weapons, perhaps restricting approaches to those less than optimal for weapon effectiveness. In other situations, defenses may increase the number of weapons that must be launched at the target. Of course, with the end of the Soviet Union, the world's most challenging air defenses will evolve more slowly. However, capable systems of the late Soviet years could well be sold to potential adversaries of the United States in regional conflicts. All of this must be seen in light of a growing sensitivity on the part of the American public to U.S. casualties in combat. The result will probably be to place a greater premium on stealthy platforms or platform-munition combinations permitting standoff launch.

- **Enemy countermeasures** can greatly alter the effectiveness of some PCS systems by decreasing the ability of intelligence to locate the target or the ability of the munitions to acquire and attack the target. Techniques can include deceptive measures, mobility, decentralization, hardening, and other techniques that influence the ability to find and damage the target. While some techniques, such as altering the array of a formation, can be applied within a tactical scenario, others, such as decentralization of command-and-control networks, might require extended periods of time and consequently occur within a strategic context.

- **Level of intelligence.** Limitations on the amount of targeting intelligence available may restrict or prohibit the practical employment of certain classes of PCS weapons in unplanned-for scenarios. This restriction applies most strongly to systems using an automatic target recognition algorithm for terminal-seeker lock-on. If target imagery in the sensor's electromagnetic band is unavailable from the desired aspect, the weapon may not be usable. Some imaging systems may be able to compensate for this shortcoming by utilizing man-in-the-loop guidance to select the final aim point, but even in this situation target intelligence must be adequate for the human operator to recognize the target. The latter restriction also applies to semiactive systems such as LGBs. The GPS/INS weapon class needs only accurate target geographical coordinates, which can be less difficult to obtain.[5]

These variables can interact with each other. Indeed, when any of the other variables are combined with collateral-damage constraints, the use of many PCS systems becomes problematic. Bad weather can delay target acquisition until too late to make an effective attack or can result in changing the weapon of choice from an LGB to a less-precise JDAM, running up the risk of collateral damage. Countermeasures such as GPS jamming can also increase a PGW's CEP to the point where the risk of collateral damage becomes too high. Poor intelligence means that the wrong target might be struck, or that the target coordinates may be sufficiently imprecise that they increase CEP to the point where collateral damage again becomes a problem.

---

[5]Geographical coordinates of the target to some level of accuracy are a requirement for almost all PCS systems. Even man-in-the-loop and anti-emitter weapons must be placed in a "basket," a position close enough to the target that the sensor can see it and the control system can successfully maneuver the weapon in.

The issue of intelligence support and mission planning deserves some additional attention because it has implications for another set of systems—those used to collect the intelligence. For effective use of PGWs, the following types of information are generally required:

- Very accurate location of the target in either absolute geodetic coordinates or relative target coordinates from a known location or from the weapon launch platform.

- Precise location of the critical aim point.

- Susceptibility of the target to functional kill (rather than structural damage).

- High-resolution imagery of the target and objects in the vicinity of the target (to support an assessment of collateral-damage potential as well as for weapon mission planning).

- Capabilities and locations of enemy defenses and other countermeasures.

- Weather at the target.

- Accurate multisource data for damage assessment (functional kill can be difficult to assess using poststrike imagery alone).

Note that this list includes information on the other critical variables discussed above. In addition to these general requirements, any given PGW on any given mission may require further intelligence support and mission-planning aids. The various requirements are listed in Table 2.3 for each of the PGW categories given in Table 2.1. For more detail on both the general and specific requirements, see Appendix B.

Not specifically addressed in Table 2.3 is the issue of timeliness. From an operational perspective, timeliness is a key variable. For example, during a conflict, the real-time collection and dissemination of intelligence on critical mobile targets can be crucial to successfully attacking those targets. At the other extreme, for strategic targets that are part of a combatant CINC's operational plan, the timelines for the collection and processing of target data, and the production and dissemination of intelligence products to support targeting and mission planning can be measured in months.

Table 2.3

Intelligence Support Requirements for PGWs

Functional Requirements	PGW Category[a]					
	1	2	3	4	5	6
General						
Accurate target coordinates	X	X	X	X		X[b]
Critical-aim-point selection	X	X	X	X	X	
Collateral-damage prediction	X	X	X	X	X	X
Countermeasure evaluation	X	X	X	X	X	X
Weather forecast	X	X[c]	X	X		X
Battle damage assessment	X	X	X	X	X	X
Specific						
Mission rehearsal/simulation	X		X	X		
Moving-target location prediction		X[c]				X
Terrain map (TERCOM) production			X			
Scene map (DSMAC) production			X			
Target template production				X		
Emitter analysis					X	

[a]1—Man-in-the-loop (LGBs, SLAM); 2—GPS-aided INS alone (CALCM, JDAM, ATACMS, JSOW); 3—Add terminal scene-matching sensors (TLAM); 4—Add target-imaging sensors instead (TLAM-C Blk IV, improved JDAM and JSOW); 5—Anti-emitters (HARM); 6—Smart submunitions (SFW, BAT).

[b]Small-footprint weapons (e.g., SFWs) require accurate target coordinates when used against small ground targets or larger units that are widely dispersed.

[c]For submunition carriers.

Clearly, U.S. ability to employ PCS effectively and flexibly in future conflicts will depend as much on the intelligence-support framework as it will on the platforms and munitions themselves. The greatest burden is likely to come from the need to support target-imaging sensors (category 4), such as in TLAM-C Block IV, which require production of a template for each target.

For any of the scenarios in Table 2. 2, or for a variant in which one or more variables differ in value from those shown, we can assess in an approximate way the appropriateness of PCS weapons against a variety of pertinent target types. This is done for several notional scenarios in Tables 2.4 through 2.7. These tables present qualitative assessments of the appropriateness of various PGWs for attacking an array of targets. These assessments take into account weapon accuracy and lethality specifications; target size, hardness, and mobility; and contextual factors such as the sensitivity to and potential for collateral damage and the ease with which targets can be identified and (in the case of mobile targets) tracked. Assessments thus reflect the extent to which a weapon is intended for use against a target type *in the context designated.*

Rather than adhering to the scenarios in Table 2.2, we have performed the assessments for variants encompassing a broader range of constraints imposed by critical variables. The first scenario (Table 2.4) is relatively unconstrained: generally good weather, air defenses near the target area limited to hand-held systems, potential for (or tolerance of) moderate collateral damage (some important targets near civilian activities, many not). The table illustrates the wide variety of choices available for some target types, e.g., air defense installations that are relocatable but are likely to remain fixed long enough for intelligence assets to be useful in targeting them. It also illustrates PGWs' limited potential, even under the best conditions, for successfully attacking targets that move about within intelligence-cycle times and targets that are deeply buried.

The Table 2.5 scenario is like that in Table 2.4, except that the potential for collateral damage is now high (many important targets adjacent to civilian activities)—or the tolerance for such damage is low. Thus, some targets judged to be primary for standoff weapons with dumb submunitions (JSOW with CEBs, and ATACMS with APAM) in Table 2.4 are only secondary here.

In Table 2.6, a further constraint is added—that of prevailing low clouds and frequent rain. Here, the choices are dramatically curtailed. The use of weapons relying on laser or infrared guidance (Hellfire, Maverick, SLAM, and LGB) is constrained. With use of LGBs restricted to breaks in the weather, there is no longer a reliable knockout punch against buried targets, and the onus for attacking softer targets falls on a narrower set of options.

For Table 2.7, a final critical variable is brought into play: Air defenses are strong to begin with and cannot be suppressed before the target types listed must be attacked. In this case, options are limited to standoff weapons such as JSOW, TLAM, JASSM, and ATACMS with BAT (assuming the kind of low tolerance for attrition that is likely to prevail in such regional conflicts). Even some of these, e.g., JSOW,

Table 2.4

Applicability of Different PCS Weapons for Various Targets in a Good-Weather, Weak-Air-Defense, Medium-Potential-Collateral-Damage Scenario[a]

Weapon	Hard Fixed				Soft Fixed		Mobile		
	Bunkers	Structures	Buried	Deep	Point	Area	Relocatable	Large/Armored	Small[b]
Hellfire	Not	Not	Not	Not	Not	Not	Primary	Primary	Second
Maverick	Second	Not	Not	Not	Second	Not	Primary	Primary	Second
LGB (500)	Not	Primary	Not	Not	Primary	Second	Primary	Not	Second
LGB (1000)	Second	Primary	Second	Not	Primary	Second	Primary	Not	Second
LGB (2000)	Primary	Primary	Primary[c]	Second[c]	Primary	Second	Primary	Not	Second
JDAM	Second	Primary	Second	Not	Primary	Primary	Primary	Not	Not
JDAM PIP (2000)	Primary	Primary	Second	Not	Primary	Primary	Primary	Not	Second
JSOW/CEB	Not	Not	Not	Not	Second	Primary	Primary	Second	Not
JSOW/SFW	Not	Not	Not	Not	Not	Not	Second	Primary	Not
JSOW P3I	Not	Primary	Not	Not	Primary	Second	Primary	Not	Second
WCMD/CEB	Not	Not	Not	Not	Second	Primary	Primary	Second	Not
WCMD/SFW	Not	Not	Not	Not	Not	Not	Second	Primary	Not
SLAM	Not	Second	Not	Not	Primary	Second	Primary	Not	Second
SLAM-ER	Second	Second	Not	Not	Primary	Second	Primary	Not	Second
TLAM,[d] JASSM	Second	Primary	Not	Not	Primary	Second	Primary	Not	Not
ATACMS/APAM	Not	Not	Not	Not	Second	Primary	Primary	Second	Not
ATACMS/BAT	Not	Not	Not	Not	Not	Not	Second	Primary	Not
ATACMS/BAT P3I	Not	Not	Not	Not	Not	Not	Primary	Primary	Second

Table 2.4—continued

[a]Weather not limiting, medium-altitude profile to avoid hand-held air defenses near the target area, collateral-damage potential similar to ODS.

[b]PCS weapons are effective against small mobile targets; they are characterized here as "secondary" because of the difficulty in locating and attacking the target before it moves.

[c]GBU-28s are LGBs specifically intended for buried and deep targets, but they are somewhat limited in number.

[d]Unitary.

Primary:	Intended target class for weapon system.
Second:	Possible secondary target for weapon system; may require large number of weapons or involve uncertainties in performance.
Not:	Weapon system not appropriate for target; target invulnerable to warhead or requires excessive number of weapons.

Hard, fixed targets include weapon storage (bunkers), tank-production plants (structures), mid-level command posts (structures), and national command posts (deep).

Soft, fixed targets include vulnerable points of buildings (point) and supply dumps or POL facilities (area).

Mobile targets include air defense units (relocatable), armored divisions (large), and transporter-erector-launchers (small). Relocatable targets do not move as often as other mobile targets and can thus be targeted within the intelligence-cycle time if they can be located; but, unlike fixed targets, they cannot be usefully identified before C-day, the day the decision is made to deploy military combat forces.

NOTE: The table is meant to be read down the columns in search of one or more favorable ratings. It is not meant to support comparative evaluations of weapon system versatility across target types, which would have to take account of other factors.

Table 2.5

Applicability of Different PCS Weapons for Various Targets in a Good-Weather, Weak-Air-Defense, High-Potential-Collateral-Damage Scenario[a]

Weapon	Hard Fixed				Soft Fixed		Relocatable	Mobile	
	Bunkers	Structures	Buried	Deep	Point	Area		Large/Armored	Small[b]
Hellfire	Not	Not	Not	Not	Not	Not	Primary	Primary	Second
Maverick	Second	Not	Not	Not	Second	Not	Primary	Primary	Second
LGB (500)	Not	Primary	Not	Not	Primary	Second	Primary	Not	Second
LGB (1000)	Second	Primary	Second	Not	Primary	Second	Primary	Not	Second
LGB (2000)	Primary	Primary	Primary[c]	Second[c]	Primary	Second	Primary	Not	Second
JDAM	Second	Primary	Second	Not	Primary	Primary	Primary	Not	Not
JDAM PIP (2000)	Primary	Primary	Second	Not	Primary	Primary	Primary	Not	Second
JSOW/CEB	Not	Not	Not	Not	Second	Second	Second	Second	Not
JSOW/SFW	Not	Not	Not	Not	Not	Not	Second	Primary	Not
JSOW P3I	Not	Primary	Not	Not	Primary	Second	Primary	Not	Second
WCMD/CEB	Not	Not	Not	Not	Second	Primary	Primary	Second	Not
WCMD/SFW	Not	Not	Not	Not	Not	Not	Second	Primary	Not
SLAM	Not	Second	Not	Not	Primary	Second	Primary	Not	Second
SLAM-ER	Second	Second	Not	Not	Primary	Second	Primary	Not	Second
TLAM,[d] JASSM	Second	Primary	Not	Not	Primary	Second	Primary	Not	Not
ATACMS/APAM	Not	Not	Not	Not	Second	Second	Second	Second	Not
ATACMS/BAT	Not	Not	Not	Not	Not	Not	Second	Primary	Not
ATACMS /BAT P3I	Not	Not	Not	Not	Not	Not	Primary	Primary	Second

Table 2.5—continued

[a] Weather not limiting, medium-altitude profile to avoid hand-held air defenses near the target area, many targets interspersed with civilian structures.

[b] PCS weapons are effective against small mobile targets; they are characterized here as "secondary" because of the difficulty in locating and attacking the target before it moves.

[c] GBU-28s are LGBs specifically intended for buried and deep targets, but they are somewhat limited in number.

[d] Unitary.

Primary:	Intended target class for weapon system.
Second:	Possible secondary target for weapon system; may require large number of weapons or involve uncertainties in performance.
Not:	Weapon system not appropriate for target; target invulnerable to warhead or requires excessive number of weapons.

Hard, fixed targets include weapon storage (bunkers), tank-production plants (structures), mid-level command posts (buried), and national command posts (deep).

Soft, fixed targets include vulnerable points of buildings (point) and supply dumps or POL facilities (area).

Mobile targets include air defense units (relocatable), armored divisions (large), and transporter-erector-launchers (small). Relocatable targets do not move as often as other mobile targets and thus can be targeted within the intelligence-cycle time if they can be found; but, unlike fixed targets, they cannot be usefully identified before C-day, the day the decision is made to deploy military combat forces.

NOTE: The table is meant to be read down the columns in search of one or more favorable ratings. It is not meant to support comparative evaluations of weapon system versatility across target types, which would have to take account of other factors.

Table 2.6

Applicability of Different PCS Weapons for Various Targets in a Bad-Weather, Weak-Air-Defense, High-Potential-Collateral-Damage Scenario[a]

Weapon	Hard Fixed				Soft Fixed		Relocatable	Mobile	
	Bunkers	Structures	Buried	Deep	Point	Area		Large/Armored	Small[b]
Hellfire	Not	Not	Not	Not	Not	Not	Second	Second	Second
Maverick	Second	Not	Not	Not	Second	Not	Second	Second	Second
LGB (500)	Not	Second	Not	Not	Second	Second	Second	Not	Second
LGB (1000)	Second	Second	Second	Not	Second	Second	Second	Not	Second
LGB (2000)	Second	Second	Second[c]	Second[c]	Second	Second	Second	Not	Second
JDAM	Second	Primary	Second	Not	Primary	Primary	Primary	Not	Not
JDAM PIP (2000)[d]	Primary	Primary	Second	Not	Primary	Primary	Primary	Not	Second
JSOW/CEB	Not	Not	Not	Not	Second	Second	Second	Second	Not
JSOW/SFW[e]	Not	Not	Not	Not	Not	Not	Second	Primary	Not
JSOW P3I[d]	Not	Primary	Not	Not	Primary	Second	Primary	Not	Second
WCMD/CEB	Not	Not	Not	Not	Second	Primary	Primary	Second	Not
WCMD/SFW[e]	Not	Not	Not	Not	Not	Not	Second	Primary	Not
SLAM	Not	Second	Not	Not	Second	Second	Second	Not	Second
SLAM-ER	Second	Second	Not	Not	Second	Second	Second	Not	Second
TLAM,[d,f] JASSM[d]	Second	Primary	Not	Not	Primary	Second	Primary	Not	Not
ATACMS/APAM	Not	Not	Not	Not	Second	Second	Second	Second	Not
ATACMS/BAT[e]	Not	Not	Not	Not	Not	Not	Second	Primary	Not
ATACMS/BAT P3I	Not	Not	Not	Not	Not	Not	Primary	Primary	Second

Table 2.6—continued

[a]Low clouds and rain prevail, medium-altitude profile to avoid hand-held air defenses near the target area, many targets interspersed with civilian structures.

[b]PCS weapons are effective against small mobile targets; they are characterized here as "secondary" because of the difficulty in locating and attacking the target before it moves.

[c]GBU-28s are LGBs specifically intended for buried and deep targets, but they are somewhat limited in number.

[d]"Primary" entries assume that this weapon will be equipped with a sensor or other accuracy-enhancing device that will work in adverse weather.

[e]"Primary" entries assume that the sensor on this weapon, although affected by the weather, can work with the short acquisition ranges below a cloud cover.

[f]Unitary.

Primary:	Intended target class for weapon system.
Second:	Possible secondary target for weapon system; may require large number of weapons or involve uncertainties in performance.
Not:	Weapon system not appropriate for target; target invulnerable to warhead or requires excessive number of weapons.

Hard, fixed targets include weapon storage (bunkers), tank-production plants (structures), mid-level command posts (buried), and national command posts (deep).

Soft, fixed targets include vulnerable points of buildings (point) and supply dumps or POL facilities (area).

Mobile targets include air defense units (relocatable), armored divisions (large), and transporter-erector-launchers (small). Relocatable targets do not move as often as other mobile targets and thus can be targeted within the intelligence-cycle time if they can be located; but, unlike fixed targets, they cannot be usefully identified before C-day, the day the decision is made to deploy military combat forces.

NOTE: The table is meant to be read down the columns in search of one or more favorable ratings. It is not meant to support comparative evaluations of weapon system versatility across target types, which would have to take account of other factors.

Table 2.7

Applicability of Different PCS Weapons for Various Targets in a Bad-Weather, Strong-Air-Defense, High-Potential-Collateral-Damage Scenario[a]

Weapon	Hard Fixed				Soft Fixed		Mobile		
	Bunkers	Structures	Buried	Deep	Point	Area	Relocatable	Large/Armored	Small[b]
Hellfire	Not	Not	Not	Not	Not	Not	Second	Second	Second
Maverick	Second	Not	Not	Not	Second	Not	Second	Second	Second
LGB (500)	Not	Second	Not	Not	Second	Second	Second	Not	Second
LGB (1000)	Second	Second	Second	Not	Second	Second	Second	Not	Second
LGB (2000)	Second	Second	Second[c]	Second[c]	Second	Second	Second	Not	Second
JDAM	Second	Second	Second	Not	Second	Second	Second	Not	Not
JDAM PIP (2000)	Second	Second	Second	Not	Second	Second	Second	Not	Second
JSOW/CEB	Not	Not	Not	Not	Second	Second	Second	Second	Not
JSOW/SFW[d]	Not	Not	Not	Not	Not	Not	Second	Primary	Not
JSOW P3I[e]	Not	Primary	Not	Not	Primary	Second	Primary	Not	Second
WCMD/CEB	Not	Not	Not	Not	Second	Second	Second	Second	Not
WCMD/SFW	Not	Not	Not	Not	Not	Not	Second	Second	Not
SLAM	Not	Second	Not	Not	Second	Second	Second	Not	Second
SLAM-ER	Second	Second	Not	Not	Second	Second	Second	Not	Second
TLAM,[d,f] JASSM[e]	Second	Primary	Not	Not	Primary	Second	Primary	Not	Not
ATACMS/APAM	Not	Not	Not	Not	Second	Second	Second	Primary	Not
ATACMS/BAT[d]	Not	Not	Not	Not	Not	Not	Second	Primary	Not
ATACMS /BAT P3I	Not	Not	Not	Not	Not	Not	Primary	Primary	Second

Table 2.7—continued

[a]Low clouds and rain prevail, air defenses limit approach and overflight, many targets interspersed with civilian structures.

[b]PCS weapons are effective against small mobile targets; they are characterized here as "secondary" because of the difficulty in locating and attacking the target before it moves.

[c]GBU-28s are LGBs specifically intended for buried and deep targets, but they are somewhat limited in number.

[d]"Primary" entries assume that the sensor on this weapon, although affected by the weather, can work with the short acquisition ranges below a cloud cover.

[e]"Primary" entries assume that this weapon will be equipped with a sensor or other accuracy-enhancing device that will work in adverse weather.

[f]Unitary.

Primary:	Intended target class for weapon system.
Second:	Possible secondary target for weapon system; may require large number of weapons or involve uncertainties in performance.
Not:	Weapon system not appropriate for target; target invulnerable to warhead or requires excessive number of weapons.

Hard, fixed targets include weapon storage (bunkers), tank-production plants (structures), mid-level command posts (buried), and national command posts (deep).

Soft, fixed targets include vulnerable points of buildings (point) and supply dumps or POL facilities (area).

Mobile targets include air defense units (relocatable), armored divisions (large), and transporter-erector-launchers (small). Relocatable targets do not move as often as other mobile targets and thus can be targeted within the intelligence-cycle time if they can be located; but, unlike fixed targets, they cannot be usefully identified before C-day, the day the decision is made to deploy military combat forces.

NOTE: The table is meant to be read down the columns in search of one or more favorable ratings. It is not meant to support comparative evaluations of weapon system versatility across target types, which would have to take account of other factors.

ATACMS, could be compromised in their applicability if U.S. forces have not yet established air superiority.

Table 2.7 assumes stealth aircraft are not employed. If they are employed, the applicability of PCS weapons will look more like that given in the preceding tables, limited by the weapon-carrying capability of the particular stealth aircraft. For example, the buried targets will still not be primary for GBU-28s, because they cannot be carried by existing stealth aircraft.

PCS POTENTIAL FOR ACHIEVING CAMPAIGN OBJECTIVES

The analyses presented so far furnish the basis for a first-order assessment of PCS potential for accomplishing tasks to achieve various campaign objectives. Analyses such as those in Tables 2.4 through 2.7 allow us to determine whether there is a PCS system that has the potential to attack targets of the type required for a particular military task. Comparing such tables for different scenarios allows conclusions to be drawn about the robustness of the capability. Finally, other factors need to be taken into account, e.g., available number of weapons of a particular type or number of delivery platforms required.

Here we present a summary assessment of PCS potential for achieving a representative range of military tasks. When we consider the objectives to which PGWs were intended to contribute, we find their potential is generally high. Our assessments, nonetheless, point out some shortcomings—not surprising, given the sensitivities already identified.[6] When we consider objectives for which PCS weapons were not originally intended, we find their potential to be quite limited at present.

We emphasize that the following assessments are based on a mix of qualitative and quantitative methodologies informed by RAND's

---

[6]Some shortcomings have already been documented. Despite the overwhelming success associated with the Persian Gulf War, after-action reports highlight many significant and some unexpected areas of PCS shortcomings. Many missions were canceled because of adverse weather. In several cases, munitions were unable to penetrate and destroy deeply buried targets. Finally, terminally guided cruise missiles and other PGWs did produce collateral damage.

broad experience in analyzing military operations. In addition, we benefited from the advice of knowledgeable representatives of the four Services. (The judgments arrived at, however, are our own.) We did not attempt within the limited scope of our study to rank the benefits or assess the costs of the potential solution directions we considered. Such an analysis could be helpful in refining our conclusions.

TASK: Attack Fixed or Relocatable Enemy Air Defenses

PCS systems are effective at attacking soft installations at air defense sites and, if equipped with penetration capability, are able to attack some moderately hard installations. The capability against fixed installations is good, but many of the sites of interest are relocatable, perhaps within the intelligence-cycle time. Capability against those relocatable targets is limited.[7] Prompt target identification and mission planning are critical in this role. Extreme weather that hinders terminal-sensor effectiveness can reduce the capability to effectively employ unitary weapons; those employing submunitions are less affected because of their broader area of coverage.

TASK: Destroy or Damage Aircraft in the Open or in Revetments

PCS systems can be very effective in this role when employing submunitions to provide area coverage. In general, less precision is required in this role, and weather has less effect on delivery (though it can influence the munitions' dispersion pattern somewhat). Long-range systems such as TLAM-D can be very effective in this role, since they can attack multiple discrete aim points without exposing a carrying aircraft to defenses around the base. However, the combination of mission-planning difficulties and the small numbers of standoff weapons available can make effective use of PCS in this role somewhat difficult.

---

[7]In Table 2.4, we list these as primary target types for most of the weapons listed, but success against these targets requires that they be found—and that can be challenging. HARM has capabilities against relocatable targets that are emitting.

TASK: Destroy or Damage Aircraft in Hardened Shelters

Some PCS systems have little capability against very hard shelters because of their nonpenetrating warheads, but others, such as LGBs with hardened warheads, are effective. However, the carrying aircraft have to get close enough to the target, so air defenses en route and at the target must have been suppressed (unless the aircraft are stealthy). Improved warheads will make systems such as TLAM much more capable against this kind of target. Because of the target's hardness, accuracy can be very important in striking the target's vulnerable point. Consequently, weather disruptions to terminal guidance can be a factor in the success of the mission.

TASK: Disrupt Electric-Power, Defense, and Fuel-and-Lubricant Production[8]

PCS systems are generally good at this mission, provided that adequate mission-planning and intelligence-preparation time is available. PGWs are particularly good at selectively attacking moderately hard critical nodes inside an installation. However, very bad weather can limit the accuracy of systems to the point where their use may not be desirable. Collateral-damage constraints can be very serious for this type of attack, because many targets are located in or near cities with sensitive facilities, and this may limit the applicability of PCS systems. System failures, failure to lock onto the proper target, and even enemy counterfire are just a few of the factors that can lead to unanticipated damage in urban areas. Improved terminal sensors could decrease the impact of bad weather; improved use of man-in-the-loop capability could allow for final target confirmation.

TASK: Destroy Moving Armor

Stopping an advancing army is largely a function of destroying its lead elements and cutting off and then destroying follow-on eche-

---

[8]These tasks are similar enough in terms of their amenability to PCS to be assessed as one.

lons and resupply. We assess current capabilities as limited.[9] Primary constraints are insufficient stocks of advanced submunitions, weather limitations, exposure to terminal defenses, and intelligence support.[10] It appears that future capability will be substantially improved. Primary enhancements will be ATACMS with BAT, WCMD/SFW, and JSOW carrying SFW submunitions. Key limitations will likely be quantities of deliverable weapons and exposure to terminal defenses (particularly for WCMD/SFW). Also, limitations on the collection, interpretation, and communication of intelligence information will probably continue to constrain the effectiveness of precision strike against moving targets for the foreseeable future.[11] Because in many cases destruction of moving armor will be among the key objectives to be achieved in the very first days of a conflict, a high priority should be placed on overcoming the limitations to PCS use for this task. (Here and elsewhere in this discussion, it must also be borne in mind that the capabilities of some future systems have yet to be fully demonstrated. Thus, the future limitations we report here, which are based on performance goals, should be regarded as lower bounds on the constraints to effective use of these weapons.)

TASK: Destroy Halted Armor

Current capability is limited, although under favorable conditions, e.g., poor-quality terminal air defenses, as in Desert Storm, effective use is possible. Primary constraints are the same as for attacking moving armor. Future capability was judged improved but still limited. Primary improvements are ATACMS with BAT P3I, JSOW with SFW submunitions, and WCMD/SFW. The principal future limitation appears to be insufficient quantities of deliverable weapons; air-

---

[9]The short-range air-to-surface Maverick was a very effective antitank weapon in Desert Storm, but its utility is limited in the face of highly capable terminal air defenses where aircraft survivability is a concern.

[10]Although systems such as JSTARS represent a quantum leap in the ability to find and track moving targets, response times still need to be short enough to put weapons on targets while they can still be tracked. Also, of course, JSTARS cannot always be present or effective.

[11]This presumes that the future battlefield will be nonlinear. On such a battlefield, early availability of reconnaissance assets, i.e., deployment and defense suppression, will be constrained.

lift and prepositioning adequate for timely availability of weapons in theater may also be problematic.

TASK: Damage Bridges to Slow Invading Armies

PCS systems are generally good at this mission, provided they possess a relatively large, hardened unitary warhead ($\geq$1,000 lb), have sufficient accuracy to strike the vulnerable region of the bridge, and sufficient intelligence is available on primary and alternate routes enemy forces may take so that other paths of advance can be closed off. Some bridge types are more resistant to damage and possess very small vulnerable areas. These targets represent a challenge for PGWs. Other limitations are similar to those stated above for disrupting electric-power production. Note, however, that in general, PCS systems intended for missions against bridges have very good accuracy.

TASK: Destroy Hardened Bunkers and Deeply Buried Facilities

PCS capabilities against bunkers are limited by weather, because the primary weapons for this role are laser-guided bombs. Ability to attack buried facilities is also limited by weather and, additionally, by the unavailability of enough penetrating warheads (e.g., GBU-28s). If terminal defenses are ineffective, multiple hits by LGBs with limited penetration capability can kill the target.

TASK: Destroy Small and Very Small Mobile Targets

A number of current and future systems have sufficient accuracy and lethality to destroy mobile SAM batteries, surface-to-surface missile launchers, command posts, and other small, mobile targets. Here, limitations lie in the ability to identify, locate, and track the target with sufficient accuracy, and for sufficient duration, to deliver a weapon onto it.

TASK: Provide Long-Range Supporting Fires

Ground warfare doctrine has been evolving in such a way that a well-defined forward line of troops may not exist in the next conflict involving massed forces on the ground. Fast-moving operational ground maneuver units may press far forward into what has traditionally been regarded as deep-attack territory, striving toward an objective more important than occupation of the territory left behind. One recent example of this was in Desert Storm, in which the 101st Airborne conducted air assaults (helicopter-supported) in excess of 100 km into enemy territory.

PGWs have often been associated with deep attack, i.e., operations well beyond the proximity of friendly troops. However, this association is not a necessary one. The current ability to execute this task is indeed very constrained, principally by the potential for fratricide. However, new sensor and processing technologies (e.g., identification, friend or foe [IFF] logic) may have the capability to provide an improved third dimension for the nonlinear ground maneuvers just mentioned. For example, antiarmor submunitions with multimode sensors (e.g., BAT P3I) that are brilliant (have the ability to identify target types) may have utility in supporting deep ground attack operations. Such a capability may be particularly relevant to U.S. infantry-based operations, which often require antiarmor support. However, because collateral damage in such a task may be in the form of casualties to friendly forces, this task appears to be even more challenging than deep interdiction of moving armor (assessed above).

A Note on Force Structure

Assessments of potential for contributing to various strategies lead naturally to the question of how many of which weapons should be in the inventory. Because each PGW is best suited to a particular set of contexts, decisions regarding proper inventories of various PGWs would have to depend on projections of how often different contexts are likely to prevail, or how important they will be. This assessment includes not only factors inherent in the scenario—weather, collateral-damage constraints, etc.—but also factors related to how the scenario plays out between the United States and its opponents.

It is not sufficient to have an adequate set of choices among weapons that may take weeks to deploy to the theater in sufficient numbers. It may be necessary to commence strike operations much sooner after U.S. forces are mobilized. For such cases, it is especially important for DoD to ensure that it invests in adequate stocks of those advanced munitions that will be most needed in the crucial opening weeks of a conflict. It is also important that these stocks be positioned so that they can be delivered rapidly to forward-deployed forces or so that they can be delivered by platforms (e.g., long-range bombers or carrier-based aircraft) that require less forward-deployed support and can respond rapidly to an emerging contingency.

Chapter Three

EVOLUTION IN TECHNOLOGY AND OPERATIONS: POTENTIAL AND LIMITS

What can be done to address the shortcomings in PGW effectiveness discussed in the previous chapter? What technological opportunities may present themselves in coming years? Should defense policy-makers be exploring different directions for manufacture of PGWs and different concepts for their operation? In this chapter, we lay out some possible future paths for development and use of PCS systems and some limitations that must be kept in mind.

CORRECTING SHORTCOMINGS

As pointed out in Chapter Two, current PCS systems cannot destroy deeply buried targets or effectively attack moving armor with im-punity. What lines of research and development might be profitably pursued to overcome the barriers for PGW effectiveness in attacking such targets?

A principal constraint in attacking deeply buried targets is the lack of penetrating warheads delivered by standoff weapons. Current standoff weapons neither hit the ground fast enough nor have a shape designed for earth penetration. A possible solution is to develop rocket-powered weapons equipped with GPS-aided INS guidance and carrying depleted-uranium-rod warheads designed to penetrate to deep targets. ICBMs and SLBMs have even been consid-ered as possible buses to deliver kinetic-energy penetrators with very high velocity. However, even if penetrating weapons can be devel-oped, the defense may always have the capability to bury its targets deeper than the deepest-penetrating weapon in the U.S. inventory; hence, developing better penetrators could be a losing proposition.

In addition, it is now very difficult for the intelligence community to collect information about such hard targets; in general, the deeper they are, the harder the collection task becomes. In the short term, DoD's best option may be the continued use of laser-guided GBU-28s to attack known buried targets. However, since this weapon must be delivered over the target, it will be necessary to provide defense suppression for the nonstealthy platforms that carry it.

The ability of PCS to halt moving armor is limited by several factors, including weapon shortages and the need to fly in close to enemy formations. If BAT and WCMD/SFW perform as advertised and are provided in sufficient numbers, they would help resolve these problems. However, BAT's costliness may limit its employment. Also, BAT is to be carried by ATACMS launched from ground-based MLRS; in the early phases of a conflict, there may not be enough MLRS batteries available on the battlefield. Also, use of PCS ahead of advancing U.S. troops against targets such as mobile C3I posts may be limited by weapon availability and concerns over accuracy and friendly fire.

TECHNOLOGICAL OPPORTUNITIES: QUALITY *AND* QUANTITY?

So far, we have focused on the current strategic vision of PCS systems as force multipliers for which the same basic types and organization of force structures would be retained. In this role, PCS helps existing platforms and organizations to better perform their missions. (A good example of this is JDAM and LGBs, which significantly improve the effectiveness of tactical aviation.) Another vision, however, is possible: PCS might be used in lieu of existing forces. Over a long transition process, the strengths of today's PGWs might be combined with advances from the information revolution supporting more flexible C3 and data processing. This fusion would include partitioning functions among manned systems, hybrid systems exploiting telepresence and data links, and purely autonomous weapons.[1] It

---

[1]As the price/performance ratio of sensor and computer technology continues to decline, opportunities may eventually arise to rely increasingly on cheap, disposable robots to replace or augment manpower. It is possible, for example, to imagine "flying" a reasonably stealthy, reusable, unmanned delivery platform—a cross be-

might lead to a strategy that allows the United States to reap some of the advantages of weapon numbers, but with a smaller force.

The long-term strategic role of PCS weapons should have implications for nearer-term development options. Development of PGWs can be approached from two directions. The first focuses on producing and employing a small number of high-quality systems; the second emphasizes quantity. Resource constraints have often led the United States to pair smart platforms with dumb weapons, or smart weapons with dumb submunitions—or to expect a few smart systems to facilitate the use of more-numerous, less-expensive assets.

This approach has been necessitated, in part, because the technology necessary for PCS systems has been relatively expensive,[2] and the demands put upon them, high. However, recent technological changes have included the advent of very powerful, low-cost microprocessors and lower-cost sensor systems. It has thus become possible to contemplate highly proliferated PGWs that will be able to achieve a large fraction of the capability of more-expensive systems at lower cost.[3]

Why would a larger number of weapons be an advantage? Larger numbers would under some circumstances allow for successfully attacking targets where the accuracy of the munitions was being degraded by weather and other factors mentioned in the previous chapter. Also, when considering the effects of defenses, a larger number of weapons (or decoys) that might absorb relatively expen-

---

tween a B-2 and a UAV, but much cheaper than the former—to a series of targets, dropping cheap PGWs on each. If cued primarily by offboard sensors, such an unmanned platform might be sufficiently inexpensive as to be "semidisposable."

[2]Our references to PCS weapons as "expensive" or "high-cost" reflect the amounts that need to be budgeted to procure them and our judgment that large numbers of weapons with such high budgetary costs are unlikely to be bought. Of course, the budgetary cost of advanced weapons must be balanced against other costs saved— e.g., by way of attrition averted or reduction in the logistics tail—not to mention the benefits only advanced weapons can yield.

[3]An alternative—and one that will probably be followed for some systems—is to spend the savings on increasing capability even further. The increased capability could be either in the weapons themselves, permitting greater accuracy from standoff, or in the platforms that carry them, permitting closer approach at constant risk.

sive defensive weapons works in favor of the attacker.[4] Ultimately, if the technology allowed for it, the United States could pursue a strategy of attrition exploiting a combination of low-cost PCS systems and the intelligence support to know where they might best be used. Finally, a greater supply of weapons would be a prerequisite for multiplying the potential of a smaller force over the long term. This approach, of course, implies something of a shift in perspective—from viewing PGWs as weapons of choice against high-value targets to embracing them as the preferred munition against a wide range of strategic, operational, and tactical targets, moving armor among them. (Of course, when collateral-damage constraints prevail, simply using a larger number of weapons can be impractical.)

Paths to a Level-of-Effort PGW: New Weapon Designs

To some extent, the United States is already pursuing a proliferated-weapon strategy with the development of JSOW and JDAM for manned aircraft. While both weapons still require relatively high-performance platforms to facilitate delivery in contested environments, this is less true of JSOW. In fact, both weapons reflect many of the qualities that may be desirable in future PCS systems: relatively low cost, some degree of standoff, and availability in large numbers. All are important attributes in realizing a level-of-effort PGW.

Realization of a true LOE weapon will have to await further advances in microprocessor and sensor systems. One near-term initiative of that kind—improved automatic target recognition—is discussed in Appendix C. It should also be possible now to undertake various system development strategies that might bring LOE PGWs closer to reality. One such path is to produce a common bus with alternative payloads that can be rapidly changed to meet operational requirements. For example, it might be possible to facilitate the reconfiguration of a PGW from a unitary warhead designed to destroy struc-

---

[4]This same logic—of proliferating relatively cheap offensive weapons in the face of relatively expensive defensive ones—has great salience historically in the debates surrounding various proposed U.S. antiballistic missile defenses, from Safeguard to "Star Wars."

tures to a submunition payload capable of attacking mobile air defenses. Such an approach would replace the prevalent practice of building alternative delivery buses for different categories of payloads—a practice that has led to families of PGWs whose individual variants have required costly development programs.

A second approach is to develop more-robust, low-cost variants of man-in-the-loop systems. These systems would have less-costly bus guidance schemes that would be updated by controllers en route and in the target area. The controllers could be air-, ground-, or space-based. The challenge to achieving this capability is to develop robust long-range data links at reasonable cost. By "reasonable," we mean that the cost of such new links would have to be significantly less than the cost of improved autonomous PGWs of equal lethality.

Another strategy would be to develop less-costly guidance concepts for the buses of unitary and submunition-carrying weapons. There would be a great advantage, for example, to making high-quality INSs less costly so that dependence on correction by jammable GPS systems could be reduced. An alternative is to reduce the cost of target-imaging sensors. However, while some technological developments show promise in this regard, such systems require reference target information whose cost of generation may remain high.

Paths to a Level-of-Effort PGW: New Approaches to Production

Besides examining new weapon designs, perhaps just as important is examining design "requirements." Among the issues to address are whether all categories of weapons require an overall reliability of 85 percent or greater and how much "militarization" of weapon components that are commercially available is really necessary. Relaxation of very strict design requirements may permit substantial savings. So might changes in procurement regulations to encourage the application of commercial production know-how and practices to military products. The Japanese, for example, are keeping costs low in the FS-X phased-array radar partly through application of civil-sector experience with gallium arsenide technology. More generally, the Japanese do not have the complex of regulations that induces defense manufacturers to separate commercial- and military-sector production of similar products.

The prevalent manufacturing processes for standoff PGWs designed for small-rate production could be changed to accommodate large-scale production, which should reduce unit costs. This change would not only affect the PGW integrators but also all the tiers of vendors that produce individual components used in PGWs. To achieve this outcome, the government would have to commit to a large, multiyear buy program for PGWs. To date, standoff PGWs have been bought for the most part in the hundreds and low thousands.

In looking at improving PCS systems in the future, it is important to consider ways of simplifying the process of developing onboard software for PCS systems. An important element in this process is continuing efforts to abstract away the hardware in developing the software necessary for PCS systems, and perhaps even to begin to develop a common real-time operating system that could be used across future PCS systems. This abstraction allows for simpler development, easier porting of the software from previous processors to more-advanced microprocessors, and the transporting of software developed for one system to another system.

NEW OPERATIONAL CONCEPTS

Novel uses of existing assets and the development and application of new technologies often suggest new operational concepts and new doctrine. Although the concepts discussed below often entail modifications to existing assets as well as new developments (C4I systems, delivery vehicles, warheads, etc.), we focus on the operational concept rather than the weapon.

In-Flight Replanning

One approach that should be considered in increasing PGW effectiveness is improving the responsiveness of the platforms carrying PGWs. This is especially true of stealth aircraft, which were designed to operate almost exclusively on missions that are fully planned before takeoff. This limitation does not allow much flexibility to deviate from planned routes and targets. As a consequence, stealth aircraft were not designed to fully exploit information technologies that would enable them to change missions en route, to effectively attack moving targets, or to strike stationary, high-value targets of opportunity.

The resolution of this shortcoming presents a technological challenge. Communication capabilities must be developed that allow offboard sensors and replanning activities to provide sufficient information to the platforms to change routes and targets, without compromising their stealth characteristics. Alternatively, low-probability-of-intercept onboard sensor capabilities could be developed that would provide organic capabilities to change routes and targets. A lesser challenge is posed by the need to substantially expand onboard data-processing, -storage, and -display capabilities.

Separating Hunter and Killer Functions

With the development of C3I technologies and systems, it may be beneficial to disperse weapon system functions. One example is to uncouple the hunter function from the killer function. For example, a spotter vehicle on the ground, benefiting perhaps from reduced-signature technology, finds targets and reports their locations, and a linked ATACMS responds.[5] Another concept is to use cheap acoustic sensors that are widely distributed (air-delivered or hand-placed) over the battlefield for situational awareness and targeting. If systems can be developed to permit IFF and target update signaling directly to the oncoming weapon, the killers could be essentially providing PCS close fire support to advancing U.S. ground forces.

Separation of air-based hunters and killers may also become important in the future. Because of the scarcity of stealthy aircraft, such aircraft might be used as hunters rather than killers (where they are payload-limited compared with their conventional counterparts). Stealthy aircraft in the vicinity of the target could act as a sensor and control platform. More can be accomplished with less, because

---

[5]Depending on the scenario, the potential advantages of ground spotters over air spotters are that they have long endurance and are able to compensate for weather. On the other hand, ground spotters have shorter horizons and are less mobile and deployable than air spotters.

payload in stealthy aircraft can be traded for more fuel and longer loiter times. Also, because they do not deliver the weapons, they can remain farther from the target, enhancing their survivability. UAVs, high-altitude nonstealthy aircraft (or even space systems), or modified PCS weapons or buses might also fill the hunter function. The killers could be swarms of low-cost PGWs fired from large nonstealthy aircraft operating at standoff range (or from surface systems). Signals could pass from the hunter through ground-based controllers to airborne operators and possibly on to the PGWs for in-flight correction. Or, one or more of these steps might be skipped; in the extreme case, the hunter would update the killer PGW directly while it was on its way.

Expanding the Future Contribution of Stealth Aircraft[6]

There is a tension between operating stealthy aircraft during daylight hours, when targets such as massed ground forces on the move may be at their most vulnerable, and operating at night, when the aircraft may be most survivable. Should stealthy aircraft be confined to night operations, when they might be less effective, while less-survivable aircraft are sent to operate in the daylight?

The answer depends on the relative daytime attrition and effectiveness and the replacement, operational, and opportunity costs of the stealthy and nonstealthy aircraft. The first four factors yield a cost-per-target-killed for each type of aircraft. Opportunity costs take into account the best alternative use of the aircraft; these are likely to be higher for the stealthy aircraft, considering that it alone can accomplish certain missions against high-value targets. Thus, the process of deciding if a stealthy aircraft should operate during daylight hours is a complex one and very dependent on the context of the possible operation. Attrition, for example, could be reduced by the assembly of force packages containing support aircraft.

The one thing to keep in mind is that attrition will be experienced in operations over hostile territory, even by low-signature aircraft. Some losses of stealthy aircraft will have to be anticipated if their utility is not to be constrained. However, an entirely different ap-

---

[6]A more detailed version of this discussion is presented in Appendix C.

proach could be adopted, e.g., using other PCS systems to perform the function of manned aircraft operating in highly contested airspace.

Stealth Anti-Emitter Weapon Concept

In the 1980s, DoD supported the development of TACIT RAINBOW, a modest-loiter-time, nonstealthy anti-emitter weapon to complement existing direct anti-emitter weapons such as HARM. High unit cost, technical problems, perceived operational shortcomings, and a substantial reduction of the intended target base (former Warsaw Pact air defenses) led to the cancellation of the program. However, advances in electronics and small-engine and computer technologies may permit the development of large numbers of very low-cost, small, expendable anti-emitter drones.

Alternatively, advances in stealth, plus the other technologies, may permit the development of long-endurance, stealthy, anti-emitter weapons. This category of weapon could be employed against high-value targets such as the critical emitter of an air defense system or a critical communication node. It could be deployed over enemy territory to loiter for substantial amounts of time, awaiting emissions from specific targets before initiating attacks. If the preplanned emitters did not come up during the weapon's operating cycle, it could be designed to fly back to a recovery area.

Additional research is warranted to determine which approach or combination of approaches is most effective.

Unconventional Precision Strike[7]

The development of novel warhead concepts may permit precision strike assets to participate in nonlethal operations. Possibilities include using high-power microwave (HPM) generators to disrupt

---

[7]Another unconventional approach to precision strike—information warfare—is discussed in more detail in Appendix C.

electronics and communication devices,[8] nonlethal chemicals to disable machinery, or aerogels as a quick-acting minefield.

These weapon concepts may address some of the identified short-comings in PCS capability. For example, if a command center is buried too deeply to be attacked with conventional warheads, HPM warheads could be used to sever the center's communication links and aerogels could be used to block the entrances, thus rendering the center ineffective and preventing easy repair.

LIMITS ON POTENTIAL: ACTION AND REACTION

In the preceding discussion, we have mentioned here and there the challenges posed by enemy countermeasures. This is an important enough point to make separately. Whether PGWs can substitute for personnel, whether it makes sense to produce them in quantity, whether they can duplicate their Desert Storm successes in other scenarios—the answers to all these questions depend on how the long-term competition between PCS systems and countermeasures plays out. As mentioned above, the offensive systems may be expensive, but they may be less expensive than the defensive systems required to prevent any leakage to a critical target. On the other hand, inexpensive devices that can jam guidance signals to a battery of incoming weapons could make it prohibitively costly to attack less-critical targets. Relative resources available to the United States and its potential adversaries must also be considered.

The potential of PGWs could thus take any of the paths shown in Figure 3.1. As more-advanced enabling technologies are brought to bear, PCS potential could increase, but these improvements may be vitiated by possible enemy reactions. At any point in the future, the United States may be at a disadvantage in the long-term competition with respect to certain capabilities, and an adversary may choose that point to mount an aggressive challenge. Before undertaking a particular system development strategy, it may be wise to project the long-term measure-countermeasure competition to gain some in-

---

[8]Countermeasures such as shielding and radiation-hardened electronics may reduce the effectiveness of HPM warheads. However, such countermeasures can be costly or impracticable.

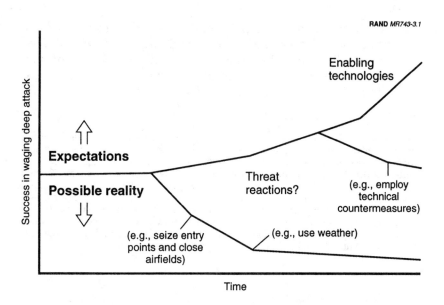

Figure 3.1—Possible Future Paths of PCS Utility

sight into how much will have to be spent to keep the United States ahead in the game an appreciable percentage of the time. And staying ahead may be challenging, because an enemy may be able to respond to U.S. technological developments, which are costly and time-consuming, with revised operational concepts, which may be cheap and quickly implemented. At a minimum, the cost and effectiveness of proposed weapon systems should be assessed against an *adaptive* adversary.

Tempering expectations may be an important step in ensuring that PCS weapons eventually do achieve whatever potential they might have. Successes in the Persian Gulf War may have led the American public and some of its elected representatives and other government officials to believe that the United States now has a tool for militarily imposing its will on an adversary where necessary. More-advanced weapons, of course, only provide a more effective way of destroying certain targets. We do not understand the implications of critical-target destruction for enemy policymaking and grand strategy.[9] It is important that those who must approve funding for PGWs and direct their use realize both the potential and the limitations of such systems. Failure to recognize the latter could lead to overestimates of the nation's ability to achieve its military objectives.

---

[9]We do not mean to suggest here that the behavior of certain adversaries is unpredictable because they are irrational. In fact, it has been argued that seemingly irrational behavior on the part of dictatorial regimes can be understood if viewed in the context of the dictator's domestic power relations, among other things (see Kenneth Watman et al., *U.S. Regional Deterrence Strategies*, MR-490-A/AF, RAND, Santa Monica, Calif., 1995). The point Watman and his coauthors make is that, regardless of the form of government, the factors coming into play following a military strike or other challenge to national security are too complex to enable the prediction of behavior with any degree of confidence. (An exception might be made for weapons of mass destruction, which the United States does not intend to use first.)

Chapter Four

A DIFFERENT APPROACH TO SYSTEM ACQUISITION

Earlier we argued for a joint approach to employing PCS systems in support of campaign objectives. In Chapter Three we postulated some revolutionary technological developments that could influence PCS system acquisition and force structure. In this chapter, we suggest that more "jointness" in system acquisition might foster the development of such revolutionary concepts as level-of-effort PGWs and their efficient incorporation into the force structure.

BACKGROUND

To promote efficiency in DoD, the Key West agreement[1] assigned specific areas of enduring responsibility to the armed services—in a sense, an assignment of "product lines": Army forces are organized to conduct operations on the land; Navy forces, to conduct operations at sea; the Air Force, to conduct operations in the air. Each Service was then to provide its own basic items of equipment (i.e., platforms) according to its assigned product line and, in turn, equip these basic delivery platforms with its own weapons and munitions.

This paradigm functioned satisfactorily through the end of the Cold War, when each Service had an uncontested claim on core military operations in a particular medium—on land, at sea, across the beach, and in the air. There was overlap at the edges, but not much.

---

[1] U.S. Joint Chiefs of Staff, *Functions of the Armed Forces and the Joint Chiefs of Staff,* April 21, 1948.

But the geographic separateness of the Cold War is largely gone. Now the focus is on joint operations in regional conflicts. From a geographic standpoint, important wartime operational areas of U.S. forces significantly overlap, which has led to "turf" battles. These battles have been exacerbated by the emphasis that interdiction has received in the doctrine for modern air and land warfare and the development of advanced technology that provides the ability to see deep and attack deep with missile or manned aircraft. Finally, the collapse of the Soviet threat has brought with it more-restricted defense budgets, which have placed a greater premium on acquisition reform. Reformers have asked whether competition might not make acquisition more efficient. While this question applies to all types of defense systems, we raise it in particular with regard to the possibility of encouraging the Services to compete for development and acquisition of PCS systems.

PURPOSEFUL COMPETITION

Would inter-Service competition for system development and acquisition have beneficial effects other than cost savings? We suspect so. Certainly, Congress in its charge to the Commission on Roles and Missions recognized that competition might be beneficial. Perhaps the most important benefit other than cost savings would be the provision of a greater range of choices at early stages of the acquisition process.

Continuing the current system implies that decisions about which Service is to provide forces that contribute capability toward a stated mission area or operational objective can be made *ex ante*. That is, they can be made before the Service presents a set of options (concepts), and certainly before it has an opportunity to demonstrate how well the proposed concept might carry out the stated objectives. *Ex ante* allocation decisionmaking precludes the possibility of having more than one concept to choose among. If the proposal from the presumed cognizant institution is weak, DoD is reluctant to encourage another Service to propose a better way to achieve the operational objective.

Even in the presence of the Key West agreement, the one-proposal-per-objective approach has not been followed exclusively. Had the United States strictly adhered to *ex ante* assignment according to missions, the Navy presumably would not have been allowed to pursue long-range ballistic missiles.[2] Intent on having a role in the mission of deterring nuclear attack on the United States, the Navy devised what clearly became by the late Cold War years the most viable solution to the operational objective of a survivable nuclear force—the submarine-launched ballistic missile. We suspect that having additional options available on a more systematic basis would enhance overall capability.

Thus, what we are proposing is to manage competition in a way that promotes informed choices among a greater variety of promising new concepts. These choices should be made on the merit of the case, unhampered by a preconceived notion of "assignments" of particular missions and roles to a particular Service. Of course, not every proposal should be funded—in fact, most probably should not be. The successful use of competition as a management tool requires the Secretary of Defense to make explicit decisions about winners and losers. Otherwise, the Department of Defense will proliferate potential solutions, squandering its resources on less-competitive ideas.

And not just the Services should be included in the shaking-out process that competition provides. The areas currently most insulated from competition are those for which the defense agencies are responsible. The defense agencies were originally created to bring greater efficiency to DoD. Immediate savings could be had from a single overhead structure for each function, and it was hoped that a central decisionmaker would more rationally allocate the available resources, eliminate excess capacity, and select the "best practice" solution. While some defense agencies are certainly successful, the growth of defense-agency resource levels relative to the DoD total is startling. Once the agencies have been created, there is little that others in DoD can do to challenge them, because they occupy a monopoly position. Such a position in a declining budget environment virtually guarantees that their relative size will grow.[3]

---

[2]The Navy could argue that the authority for organizing Polaris force elements rested on "combat operations at sea." Key West is silent as to which Service organizes "strategic nuclear missile forces."

[3]Much of this discussion draws on David Chu, "Refocusing the 'Roles and Missions' Debate," November 1994 Schulze Memorial Essay, *Marine Corps Gazette.*

THE NATURE OF THE COMPETITION[4]

For competition to work, it will be necessary to have groups that conceive and advocate new systems (weapons and submunitions) and new concepts of weapons. A central group is then needed for evaluating these new concepts and systems. That group should be composed of personnel from OSD and from all Services. But the personnel should not be "representatives" of the Services to advocate Service weapons. Rather, they would have the purpose of helping to create one road map for developing and acquiring weapons and another road map for submunitions.

The group should rigorously avoid Service lines. That is, the group should operate in a mode that promotes the idea that the following course of events would not be exceptional: Someone from Service A conceives a new concept for a submunition; some laboratory in Service B is charged with demonstrating proof of principle; in turn, Service C is charged with developing and procuring stated quantities of the submunition. The submunition is used to equip various types of weapons for all Services. The "group" is the advocate—as distinct from a particular Service being the advocate. There is a "group" position as distinct from a Service position. In this construction, then, we argue that there should be a competition of new concepts for weapons to equip the platforms at issue and for submunitions to equip those weapons.

Competition suggests some changes in process and thinking. We propose that some combination of the Chairman or Vice Chairman of the Joint Chiefs of Staff, the Office of the Under Secretary of Defense for Acquisition and Technology (USD/A&T), and the Services accomplish the following:

- Effecting separate, purposeful competitions for weapons and submunitions.
- Evaluating new concepts for both weapons and submunitions.
- Selecting concepts for demonstration.

---

[4]Our approach assumes that a robust 6.1 (basic research) and 6.2 (exploratory development) technology base is sustained by DoD.

- Getting selected concepts demonstrated quickly. This process includes both defining an approach that can be accomplished quickly and gaining timely financing.

- Deciding what concepts (weapons and submunitions separately) are to be developed and produced.

- Seeing that the concepts so selected are indeed developed and procured with prudent haste.

The thrust of such a competition proposal is to create the environment and process whereby weapons and submunitions are "purple." While types of force elements and basic items of equipment are generally Service-unique, Service lines should be unraveled when it comes to weapons and submunitions.

Establishing Requirements

The Vice Chairman, assisted by the Joint Requirements Oversight Council (JROC), should be proactive in identifying critical deficiencies and thus "opportunities" for new weapons. This means stating, in the form of a "mission need statement" (MNS), which operational objectives and operational capabilities deserve increased emphasis. These statements should be short and to the point. In the case of precision strike, there should be increased emphasis on the following:

- Halting invading armies with strike weapons

- Destroying deeply buried bunkers

- Destroying facilities attendant on weapons of mass destruction.

Based on these MNSs, the role of each Service would be to create concepts—always alert to the proactive statements by the Vice Chairman as to those operational objectives and tasks that deserve special emphasis. This proactive approach would save considerable time, because personnel in the Services would not waste their time developing a separate MNS for each proposed solution and processing the MNS so that formulation of the solution concept (already stated) can commence.[5]

In summary, according to the proposed approach, the statement of "mission need" is proactively advanced by the Chairman and Vice Chairman—and on a more-or-less continuing basis. Senior officials within the Services would have the authority (and mandate) to induce the formulation of concepts without further ado.[6]

Formulating New Concepts

New operational concepts do not automatically exist. First, there must be a process that promotes an informed review of the operational needs. Next, proposed solutions must be defined through a purposeful interaction of operators, technologists, and the intelligence community. Operators understand military operations and the operational requirements or objectives to be achieved. Technologists understand what is possible based on the enabling technologies. Representatives of the intelligence community understand how adversaries of the United States might react to changing U.S. military capabilities and how intelligence operations can enhance the effectiveness of the concepts.[7]

---

[5]The time saved may be at least 18 months. This point is underscored by examining the case of the concept of putting an IGPS guidance kit on an existing bomb. The Air Force prepared and submitted an MNS some months after the concept had already been described, evaluated, and endorsed by senior Air Force officials.

[6]This approach is about weapons; such authority might not be extended to platforms for which the costs to the government are considerably greater.

[7]These interactions must begin early in the acquisition cycle, when the higher costs of providing better intelligence support should be weighed against the lower cost of placing less-stringent requirements on the technical characteristics of the weapons, as well as the lower cost of providing less intelligence support against higher costs of more-stringent requirements. In examining such trade-offs, decisionmakers also need to consider other choices available to operators to execute the missions. For example, the overall accuracy of a postulated weapon may depend on both the accuracy of the weapon's guidance system (which is the responsibility of the weapon developer) and the accuracy of the target's location (nominally provided by the intelligence community). But increasing the accuracy of one or both elements may not be the most cost-effective way of satisfying a stringent criterion of no collateral damage. Instead, an operator might wish to employ a man-in-the-loop system.

One way of causing these interactions is through the Defense Science Board and the scientific boards that attend each of the Services. DoD federally funded research and development centers might also help. Most important, however, the Services must be involved.

Evaluating New Concepts and Selecting Those to Be Demonstrated

Once concepts are formulated, a responsive and authoritative means of evaluation is needed. The Vice Chairman, assisted by the JROC forum, should play an important role in evaluating new concepts for the purpose of deciding which ones should be selected for demonstration. Over some stated cost threshold, the final decision on whether resources (if any) should be applied to the demonstration is probably a matter for the USD/A&T's Assistant for Concept Development and the Defense Advisory Board (DAB). When the matter comes before the DAB, that body would have before it

- a thorough description, by the proposing Service, of the concept being proposed, along with a full description (and cost) of the approach to demonstrate proof of principle.

- an evaluation, by the Vice Chairman and the JROC, along with the proposing Service, of the effectiveness and relevance of this concept in the context of increasing capability to achieve critical operational objectives.

- an evaluation, sponsored by the USD/A&T's Assistant for Concept Development, of technical feasibility. Making the decision to "demonstrate" need not be prolonged. After all, at this point, the government is committing only to a demonstration, which for weapons and submunitions should not cost very much. Also, at this stage the government is committed only to seeing if the concept is feasible. The evaluators of technical feasibility need only attest to a "reasonable expectation." They are not required to guarantee the outcome.

Demonstrating Concepts

Taking a creative approach to demonstrating new concepts presents the opportunity to accomplish demonstrations quickly and at minimum expense to the government. Take, for example, the demonstration that was conducted for the Airborne Warning and Control System (AWACS). An engineer had proposed that Doppler processing of radar return signals might allow detection of a low-flying aircraft from an airborne platform, even in the presence of ground clutter. The approach to demonstrating proof of principle for this concept was to let a contract to a company skilled in electronics, but with no interest in providing the radar. The Airborne Instruments Laboratories (AIL) was selected. AIL fashioned a crude antenna outside a test-bed (a Lockheed Electra) and arranged for technical representatives from various radar companies to hook up their radars (and signal processors), whose output went to a standard display module (scope). The various radar companies (Hughes, General Electric, Westinghouse, Hazeltine, et al.) appeared on schedule to "take their ride." All radars passed proof of principle, i.e., the operators at the scopes could detect aircraft flying below the test-bed, even in the presence of ground clutter. Based on these tests, the decision was made to proceed with an acquisition program to implement the concept. From definition of approach to demonstration of proof of principle took something like 18 months.

Such an approach could be applied to demonstrating the concept of "scene-matching systems" in the family of precision strike weapons. The concept of a terminal-engagement system on gravity bombs has been thought about for several years. The concept centers on getting a bomb into some basket, whereupon the scene-matching system takes over to guide the bomb to the target with very low CEP. The scene-matching systems come with various sensors—imaging infrared or microwave, CO_2 lasers, etc.

The proposed approach for demonstrating proof of principle would be for the government to provide a test-bed on which all "ready" contractors would be given a ride. The technical representatives from the various contractors would bring "brass-board" equipment (no attempt at flight hardware). These tests should be completed and the test results analyzed in not more than a year and a half. The government would control the experiment and thus could trust the results. With these results, the government would be in a position to make an informed decision on whether or not to implement the concept.[8]

Contrast this approach with the following. The government goes through a laborious source-selection process and lets a contract to one (or two) of the contractors. The horizon is now limited to that contractor and whatever concepts and types of sensors it is proposing. But at this juncture the most pertinent question is, "Can *any* contractor make a terminal-engagement system that will work?" Selecting the contractor (later on) to engineer and produce thousands of the selected systems at an affordable price is quite a different matter—a matter to be addressed after Milestone 1.

According to the proposed approach, no time is lost in demonstrating proof of principle because of negotiating contracts. Contractors are paid for their participation by a flat fixed fee that is determined and announced by the government—a fee to mostly defray the expenses incurred in hooking up their terminal-engagement systems to the test-bed. The deliverable by the contractor is "best effort." The contractor cannot, of course, be required to succeed, because ability to succeed is unknown at this point.

To summarize, a creative approach for demonstrating proof of principle has the potential of saving time, saving money, and gaining more-relevant and -reliable information about whether the concept is technically feasible.

Deciding to Implement (Milestone 1)

The decision about whether or not to implement the concept (now demonstrated) is made by the Secretary or Deputy Secretary of Defense using the Planning, Programming, and Budgeting System. For small programs, it might be made by the USD/A&T, with assistance and advice from the DAB forum. When making this

---

[8]This demonstration approach is similar in various ways to current advanced-concept technology demonstrations. However, the latter are limited to the integration of mature technologies. An objective of the approach proposed here would be to determine whether the concept is even technically feasible, so it would need to be implemented early in the acquisition process.

decision to implement the proposed concept, the participants would have in hand

- the results of the demonstration

- an evaluation by the Vice Chairman (and each sponsoring Service) of the operational feasibility of the concept and an assessment of its relevance and worth

- an analysis of the cost of implementing the concept. Such an analysis should be sponsored by the government, with OSD's Cost Analysis Improvement Group as a central player.

Then comes the decision on whether or not to proceed with a program to acquire the systems to implement the concept. The decision to proceed rests on five principal criteria:

- Operational feasibility and relevance for joint operations

- Technical feasibility

- Logistical maintainability

- Fiscal affordability

- Political acceptability.

In the presence of a well-conceived demonstration, there is no need to conduct a demonstration/validation phase after Milestone 1. Rather, the program would then go directly into engineering and manufacturing development. (Of course, an acquisition strategy that eliminates a demonstration-and-validation phase after Milestone 1 is feasible only if the demonstration in concept development was well conceived and executed.) Then, Milestone 2, according to this construction, addresses the decision on whether to proceed with low-rate initial production.

Summary

We have described a process for introducing new weapons into the operational inventory quickly, efficiently, and in the presence of the right information at the right time to make informed decisions. Such an approach might apply to all systems in general, but surely applies to strike weapons. The DoD should plan to systematically and continually upgrade basic delivery systems with new and better weapons, and to provide P3I to these weapons, as well. This calls for

- an approach that, overall, is "purple"
- an approach that promotes timely change
- a proactive means of initiating concept formulation on the most pressing problems
- a more purposeful approach to defining and developing new concepts of weapons and submunitions
- a more systematic and timely process for selecting concepts to be demonstrated
- a readily available source of funds for financing these demonstrations
- creative approaches toward demonstrating proof of principle of selected concepts quickly and with minimum cost to the government
- an approach for quickly implementing selected concepts, i.e., acquiring the systems (weapons and submunitions) for the operational inventory.

An approach to achieving those objectives would have the following characteristics:

- There would be a more-or-less continual effort on concept formulation. This effort would be in response to statements by the Chairman or Vice Chairman about tasks and operational objectives that deserve increased and special emphasis. Once concepts are defined and evaluated, concepts worthy of demonstrating would be selected, normally by the Vice Chairman or the JROC forum.
- Once selected for demonstration, the approach would be defined and executed quickly.
- Once demonstrated and selected for implementation, the acquisition system would take over.

CAMPAIGN OBJECTIVES, OPERATIONAL OBJECTIVES, AND TASKS

To aid in identifying tasks for which precision conventional strike might be suitable, we defined a broad array of campaign objectives and subsidiary operational objectives. We then laid out the tasks whose achievement would aid in attaining the operational objectives. Assessments of PCS capability for accomplishing some interesting tasks are given in Chapter Two. The full list of tasks we considered is given here, organized by campaign and operational objective.

CAMPAIGN OBJECTIVE: GAIN AND MAINTAIN AIR SUPERIORITY

Suppress enemy sortie generation

Disable operating surfaces

Destroy/damage aircraft in the open or in revetments

Destroy/damage key support facilities

Destroy/damage aircraft in hardened shelters

Suppress enemy air defenses

Destroy/disrupt fixed SAM launchers

Destroy/disrupt mobile SAM launchers and AAA

Destroy/disrupt tracking and engagement radars

CAMPAIGN OBJECTIVE: COUNTER ENEMY LONGER-RANGE BALLISTIC MISSILES

Suppress generation of ballistic-missile launches

Destroy/damage TELs in the field and disrupt operations

Destroy/damage TELs in garrisons and assembly areas

Destroy/damage fixed missile launchers

Destroy/damage missile storage and support facilities

CAMPAIGN OBJECTIVE: DENY ENEMY POSSESSION AND USE OF WEAPONS OF MASS DESTRUCTION

Damage/deny facilities for producing and storing WMDs

Destroy/damage factories and storage sites

Block entrances to tunnels and mines

Deny enemy access to key sites

CAMPAIGN OBJECTIVE: HALT INVADING ARMIES

Delay/destroy/disrupt lead elements of invading armies

Destroy/damage armored and other vehicles on the attack

Mine key routes of advance

Suppress forward-area rearm-and-refuel points for attack helicopters

Delay/damage enemy forces and logistics in the rear

Destroy/damage armored and other vehicles in convoys and assembly areas

Destroy/damage supply stockpiles

Disrupt field logistics sites and transportation nodes

Mine roads and railroads

Destroy/damage bridges and rail yards

Block tunnels and other choke points

Provide fire support to friendly forces in close contact with the enemy

Destroy/damage armored vehicles near the line of contact

Disable dismounted troops

Destroy/suppress artillery and multiple-rocket launchers

CAMPAIGN OBJECTIVE: GAIN AND MAINTAIN SEA CONTROL OR DENIAL

Disrupt enemy surface ship operations

Sink/disable ships at sea and in port

Damage/disrupt shore support facilities

Mine ports, choke points, and anchorages

Disrupt enemy submarine operations

Sink/disable submarines in port

Damage/disrupt shore support facilities

Mine ports, choke points, and anchorages

CAMPAIGN OBJECTIVE: SUPPRESS ENEMY'S WAR-SUPPORTING INFRASTRUCTURE

Disrupt enemy POL production, storage, and distribution

Damage/disrupt refineries

Destroy/damage storage facilities

Sever POL pipelines

Disable pipeline-control facilities

Disrupt off-load terminals and transshipment points

Disrupt enemy electrical-power production and distribution

Damage/disrupt generating plants

Damage/destroy key substations and transformer yards

Cut power lines

Disable grid-control facilities

Destroy/damage known backup power sources

Disrupt enemy transportation system

Damage/disrupt airports, seaports, and transshipment points

Disrupt/destroy network control and navigation systems

Disrupt enemy defense production

Damage/destroy defense-related plants and equipment

Reduce flow of defense-related imports

CAMPAIGN OBJECTIVE: SUPPRESS WILL OF ENEMY LEADERSHIP AND FORCES

Disrupt political direction of enemy's society, economy, and war effort

Destroy/disrupt key directing organs and leadership cadres

Destroy leadership and internal-security facilities

CAMPAIGN OBJECTIVE: IMPLEMENT PEACE AGREEMENT/CEASE-FIRE

Supervise/enforce disarmament

Destroy weapons caches

Interdict shipments of arms and other contraband into and within territory

CAMPAIGN OBJECTIVE: ESTABLISH AND PROTECT SAFE AREAS

Protect safe areas from external threats

Execute punitive strikes against specific facilities, sites, or installations

CAMPAIGN OBJECTIVES: VARIOUS

Degrade command and control of enemy forces

Destroy/damage command bunkers and other critical fixed targets

Destroy/damage mobile command posts

Disrupt communications

Support special operations forces in hostile territory

Provide fire support to special-operations forces

Disrupt enemy space operations

Destroy/damage space launch facilities, command centers, surveillance and tracking stations, up- and downlinks, and storage sites

Destroy/damage mobile space surveillance and tracking facilities

INTELLIGENCE SUPPORT AND MISSION-PLANNING REQUIREMENTS FOR PRECISION CONVENTIONAL STRIKE

In this appendix, we elaborate on the information given in Chapter Two regarding intelligence support and mission-planning needs for precision-guided weapons. First, we address common intelligence support needs—the top half of Table 2.3 (repeated here as Table B.1). Then, we take up needs tailored to specific PGW categories.

COMMON INTELLIGENCE SUPPORT

In this section, we discuss the intelligence support for planning and executing missions that are common to most, if not all, of the PGW categories.[1] We describe these needs in terms of functions that are supported by intelligence data (see Table B.1).

Not specifically addressed in Table B.1 is the issue of timeliness. Timeliness is discussed below for selected functional requirements. From an operational perspective, timeliness is a key variable. For example, during a conflict, the real-time collection and dissemination of intelligence on critical mobile targets can be crucial to successfully attacking those targets. At the other extreme, for strategic targets that are part of a combatant CINC's operational plan, the timelines for the collection and processing of target data, and the production and dissemination of intelligence products to support targeting and mission planning can be measured in months.

---

[1]For more information, see Myron Hura and Gary McLeod, *Intelligence Support and Mission Planning for Autonomous Precision-Guided Weapons: Implications for Intelligence Support Plan Development*, MR-230-AF, RAND, Santa Monica, Calif., 1993.

Table B.1

Intelligence Support Requirements for PGWs

Functional Requirements	PGW Category[a]					
	1	2	3	4	5	6
General						
Accurate target coordinates	X	X	X	X		X[b]
Critical-aim-point selection	X	X	X	X	X	
Collateral-damage prediction	X	X	X	X	X	X
Countermeasure evaluation	X	X	X	X	X	X
Weather forecast	X	X[c]	X	X		X
Battle damage assessment	X	X	X	X	X	X
Specific						
Mission rehearsal/simulation	X		X	X		
Moving-target location prediction		X[c]				X
Terrain map (TERCOM) production			X			
Scene map (DSMAC) production			X			
Target template production				X		
Emitter analysis					X	

[a]1—Man-in-the-loop (LGBs, SLAM); 2—GPS-aided INS alone (CALCM, JDAM, ATACMS, JSOW); 3—Add terminal scene-matching sensors (TLAM); 4—Add target-imaging sensors instead (TLAM-C Blk IV, improved JDAM and JSOW); 5—Anti-emitters (HARM); 6—Smart submunitions (SFW, BAT).

[b]Small-footprint weapons (e.g., SFWs) require accurate target coordinates when used against small ground targets or larger units that are widely dispersed.

[c]For carriers of submunitions.

Accurate Determination of Target Coordinates

Many PGWs require the most accurate target coordinates available from the Defense Mapping Agency (DMA), usually obtained from stereo imagery. Although the remaining PGW categories do not require the same accuracy, they generally require more-accurate coordinates than can be obtained from common medium-scale to small-scale (1:250,000 scale or smaller) maps such as Joint Operations Graphics (JOGs) and Operational Navigation Charts (ONCs). Imagery is used to identify many PGW targets in the first place, and that imagery (not always stereo) can serve as a source of target coordinates. The accuracy of the target coordinates depends on the quality of the support data that accompany the imagery.

Critical-Aim-Point Selection

PGWs can be delivered against targets with very high accuracy. Such accuracy enables the operators to attack critical aim points on the target with modest-sized warheads, achieving the mission objective without causing catastrophic damage.

Critical aim points are selected by well-trained intelligence personnel, known as targeteers. After reviewing the commander's guidance and the mission objectives, the targeteer uses high-resolution imagery to identify specific elements within the target that must be attacked and estimates the level of damage to meet the mission objectives. For large target complexes such as refineries and power plants, the targeteer must understand the operations within the target complex. For example, if the objective is to stop electrical-power distribution for a short period of time, the targeteer may select the transformer yard of a power plant as a critical element rather than the generator hall, which would take substantially longer to repair. Often, targeteers rely on experts in the commercial world to assist them. The high-resolution imagery of the target or target complex is usually annotated by the targeteer to indicate the precise location of the aim points for use by the operators.

To determine the critical aim points of hardened or buried targets, intelligence personnel typically need information on the target structure (external and internal), the interior layout, and the location of critical components (such as the communications equipment for a command-and-control bunker); such data can often be very difficult to obtain, especially on short notice. Some options may include the following: Archival imagery can be reviewed to research the construction history of such targets. Sometimes, data such as blueprints, building plans, construction contracts, defector reports, or, possibly, inside photographs can be obtained; these types of data are usually provided by human-source intelligence (HUMINT).

Collateral-Damage Prediction

If the target is in a cluttered environment, especially an urban area with several nontarget buildings or objects in its vicinity, high-resolution imagery of the target area is needed not only to identify the target but also to indicate the nontarget objects as well. This information is required by operators planning the missions to select flight profiles and weapon launch points and to rehearse missions that will maximize the probability of striking the target and minimize the probability of collateral damage.

Countermeasure Evaluation

The presence of enemy countermeasures is also important to PGW mission planning and execution. Obscurants in the target area, such as smoke and fires, could affect the performance of man-in-the-loop (MITL) PGWs and autonomous PGWs with terminal sensors. The presence of GPS jammers could affect the performance of PGWs that rely on GPS; it is more critical in the target area, but it could also affect the performance of long-range weapons that rely on GPS for en route navigation.

Weather Forecast

Weather information, such as height of cloud cover, humidity, and precipitation in the target area, is important to mission planning and execution for MITL PGWs and for autonomous PGWs with terminal sensors or carrying smart submunitions. Poor weather over the target area may render these PGWs ineffective. In addition, strong winds over the target area can affect the performance of PGWs with submunitions; if winds are not accounted for during mission planning, the submunitions may be blown away from the target area following their release.

Battle Damage Assessment

The employment of PGWs places a premium on accurate and timely battle damage assessment (BDA). Typically, post–PGW-strike imagery of a hardened or buried target indicates a small-diameter hole. A PGW strike against a large building or target complex may not result in its complete destruction, either. Against such a target, strikes restricted to critical aim points will result in functional damage only. Without supporting data from signals intelligence (SIGINT) or HUMINT, it may be difficult to determine whether the desired damage level has been achieved. Thus, multisource intelligence information is often required for BDA of complex targets and hardened or buried targets.

TAILORED INTELLIGENCE SUPPORT

In the following paragraphs, we discuss the specific intelligence support needs of each PGW category.

Man-in-the-Loop PGWs

Man-in-the-loop PGWs provide operators with the potential of achieving very high delivery accuracy, often less than 3 meters CEP. Aircrews, with proper intelligence support and mission planning, can hit a particular window on a building or a very small area of a target, such as a bridge piling. Such weapons are thus particularly well suited to attacks on critical aim points.

The precise location of the aim points is usually annotated on high-resolution imagery, which the pilot can use to familiarize himself with the target. If sufficient imagery support data are available to generate a 3-D perspective of the target (by mensurating the external dimensions of the target using the high-resolution imagery), the operator can use a digital imagery workstation to assist in mission rehearsal by rotating the target image to the desired approach azimuth.

If the target can be identified visually by the aircrew during daytime operations and there are no threats in the target area, then very accurate target coordinates are not essential for LGB missions. However, for nighttime operations in which the target is acquired using onboard sensors, such as the Low-Altitude Navigation and Targeting Infrared System for Night (LANTIRN) targeting pod, the critical aim point must be very accurately known. Otherwise, the target may not appear in the narrow field of view of the sensor when it is activated. The combination of high aircraft speed, the LANTIRN pod's small field of view, and the possibility of threats in the target area preclude aircrews from conducting any substantial search to acquire the target.

MITL weapons with data links require similar intelligence support and mission planning. Depending on the field of view of the weapon and the responsiveness of the weapon to adjustments by the operator, the accuracy of target geolocation may vary somewhat. That is, the bigger the field of view of the sensor and the faster the response of the weapon to aircrew adjustments, the less stringent the requirements for geolocation.

MITL weapons have inherent technical capabilities to acquire information useful for BDA. In LGB strikes, assuming the delivery aircraft is equipped with gun camera video recorders, aircrews can record the flight of the weapon into the target. Similarly, in strikes with data-link weapons, aircrews can record or see the weapon fly into the target and see the effects of the strike. The recorded information and mission report data from aircrews can be further analyzed by BDA cells and corroborated by other data to determine the effect of PGW strikes. On the basis of these data and analyses, operational commanders can determine the need and, if necessary, the approach for subsequent restrikes.

GPS-Aided INS PGWs

Using its GPS-aided INS guidance, a category 2 PGW flies to the planned absolute target coordinates (latitude, longitude, elevation). The planned target coordinates can be derived from any of several sources of information:

- The Defense Mapping Agency's Point Positioning Data Base (PPDB), which is geocoded high-resolution stereo imagery,[2] or DMA's Points Program, a service provided to selected users

- GPS receiver readings at the target prior to the start of hostilities

- Tactical platform-derived target coordinates using an onboard sensor to locate the target relative to the platform and an onboard GPS receiver to locate the platform relative to absolute coordinates.[3]

---

[2]"Geocoded" imagery means that each pixel (picture element) of the stereo imagery has associated with it a very accurate absolute coordinate referenced to the Earth (through the WGS-84 coordinate system). The hard-copy PPDB will soon be replaced by a digital version being developed by DMA.

[3]Although "differential" GPS concepts are well known, we are considering here only the nominal guidance concept using "absolute" coordinates.

The effectiveness of GPS-aided INS PGWs is directly related to the accuracy of the target's coordinates (often called target location error, or TLE) and the accuracy of the weapon's guidance, navigation, and control (GNC) system, which is dominated by the accuracy of GPS updates.[4] JDAM, now under development, has a specification of 13 meters CEP against a horizontal target. To achieve this level of delivery accuracy, very accurate target coordinates are required.

Sometimes overlooked in discussions about mission planning of GPS-aided INS PGWs is that, before GPS target coordinates are derived, the target and its critical aim points must be identified and the target must be properly weaponeered. These functions are typically performed during the target development process or may be done in conjunction with mission planning; but they must be done.

Unlike MITL PGWs, GPS-aided INS PGWs and the other autonomous PGWs cannot rely on launch-platform gun camera video or aircrew observations to support BDA, unless operational conditions (e.g., enemy defenses) allow very-short-range deliveries. Consequently, the BDA for autonomous PGWs is likely to rely mostly on offboard sensor information.

PGWs with Scene-Matching Sensors

The intelligence support and mission planning requirements of TLAM-C, the only PGW with a scene-matching sensor, are well known. For en route navigation, Block I and II Tomahawks require TERCOM maps (digital terrain elevation maps with a specific format) built primarily by DMA from high-resolution stereo imagery of areas with sufficient terrain roughness to be uniquely recognizable. With the fielding of the Digital Imagery Workstation Suite, the Cruise Missile Support Activities (CMSAs), which support Tomahawk mission planning, will also be able to build TERCOM maps using high-resolution stereo imagery as source material. TLAM-C Block III, equipped with GPS-aided INS, does not necessarily require TERCOM maps for en route navigation.

---

[4] In each of the three dimensions, the overall weapon accuracy is given by σ, calculated from $\sigma^2 = \sigma_{TLE}^2 + \sigma_{GNC}^2$.

For terminal-area planning, the CMSAs rely on PPDBs and recent (usually days or weeks, but could be longer) high-resolution imagery to build the DSMAC scenes needed by the DSMAC guidance algorithm. The Block III missile can rely on one DSMAC scene, because en route navigation is very accurate; the earlier blocks require more DSMAC scenes. Following a position update on the last DSMAC scene, the missile inertially guides (with GPS updates if Block III) to the target.

PGWs with Target-Imaging Sensors

This category of PGW will rely on a GPS-aided INS to arrive at a planned location in the vicinity of the target for sensor turn-on. The PGW will then use its target-imaging sensor and correlation algorithm to image the target area and acquire and home on the target. With this terminal-guidance scheme, these autonomous PGWs will have the potential of achieving very high delivery accuracies, comparable to MITL PGWs.

For target acquisition, this category of PGW will require target templates—very accurately mensurated 3-D descriptions of the target and objects in the immediate vicinity of the target. These templates will be used by the correlation algorithm to identify the target and the critical aim point. The building of target templates requires high-resolution stereo imagery,[5] a workstation capable of manipulating and mensurating the imagery, and personnel well-trained in template building. The intelligence support and mission-planning infrastructure for autonomous PGWs with target-imaging sensors has not been fully developed. However, all other things considered, this category of PGW will likely place the greatest burden on intelligence support and mission-planning. In that case, this weapon may be used primarily against high-value planned targets rather than emergent targets.[6]

---

[5]Depending on the type of target-imaging sensor, radar, infrared, or visible target imagery (or combinations of imagery) may be required.

[6]Because these weapon systems are often very expensive, high-value, well-defended targets requiring very accurate weapon delivery would be the most likely choices.

Anti-Emitter PGWs

These PGWs are programmed to fly out on specified azimuths to search, detect, acquire, and home on a particular electronic emitter. Aircrews planning anti-emitter PGW missions require the location, frequency, pulse width, and other characteristics of the target emitter and other emitters in the weapon's designated search area. With this information, aircrews can select the launch basket, flight profile, and search area that will maximize the probabilities that the weapon will detect, acquire, and home on the target emitter.

PGWs with Submunitions

These PGWs use the missile as a "bus" to deliver several possible types of submunitions to a specific location. (In Table B.1 we single out weapons carrying "smart" submunitions (category 6); "dumb" submunitions are associated with the delivery vehicle in the other categories.) Submunition-carrying PGWs have special intelligence support and mission-planning needs distinct enough from unitary-warhead PGWs to merit discussion beyond that related to the buses themselves.

The delivery buses may be of categories 2 or 3 (examples are ATACMS, JSOW, and TLAM-D). The submunitions may range from dumb (non-sensor-equipped) Combined-Effects Bomblets (CEBs), such as those carried by TLAM-D, to brilliant submunitions, such as the Brilliant Antiarmor Submunition (BAT) now being developed for ATACMS. BAT will include sensors and associated algorithms to detect, acquire, and home on targets. Equally important, different combinations of buses and submunitions provide different footprints, i.e., different submunition strike patterns on the ground.

In general, these PGWs can be used on fixed, soft targets such as air defenses or against mobile targets. Against fixed point or area targets, the more accurate the geolocation of the target, the better the probability of delivering the submunition to a dispense point that will maximize the probability of dumb submunitions hitting the target or the probability of smart submunitions detecting, acquiring, and homing on the target. PPDBs or the DMA Points Program can be used to derive very accurate target geolocation. Tactical targeting platforms may provide very accurate relative target coordinates to support planning and employment of PGWs with submunitions against fixed targets. Obviously, for fixed area targets, information about the distribution of objects of interest is important.

Information on the direction and speed of movement is essential for the effective employment of autonomous standoff PGWs with submunitions against moving targets. This information is needed to determine the submunition dispense point that will result in a footprint that will maximize the probability of submunitions striking the moving targets. This type of information is best provided by tactical targeting platforms such as the Joint Surveillance [and] Target Attack Radar System (JSTARS) or unmanned aerial vehicles.

Finally, weather data can be very important for effective employment of PGWs with submunitions. If the submunition has a smart sensor, very poor weather conditions may preclude target acquisition. Also, if high winds are in the target area, the submunition pattern may drift away from the target or centroid of targets unless the submunitions are delivered at very low altitude or by a guided bus.

SUPPLEMENT ON NEW TECHNOLOGIES AND CONCEPTS OF OPERATION

Here we provide full discussions of topics mentioned briefly in Chapter Three:

- Improved automatic target recognition

- Improved ground vehicle stealth

- Expanding the contribution of stealth aircraft

- Unconventional "precision strike" (information warfare).

IMPROVED AUTOMATIC TARGET RECOGNITION

PGWs with target-imaging sensors now under development (such as TLAM Block IV) will use an onboard automatic target recognition (ATR) algorithm to acquire and home on their targets with very high delivery accuracy. When such a weapon approaches the target, the weapon will activate its sensor and image the target area. The algorithm will then locate the target in the image[1] and identify the critical aim point by correlating with a preplanned target template. The target as well as a number of nearby "contextual" objects will be used to develop a unique template to avoid acquisition of a false target.

A different template must be developed for each approach azimuth and, for some sensors, for different times of day or seasons of the

---

[1]If the target is not in the image, the mission will fail because these weapons cannot search the target area. It is critical that the sensor's range and field of view are sized to compensate for errors associated with en route navigation and target location.

year. Therefore, this type of ATR algorithm will be limited to fixed targets and, because each target and its nearby objects will be unique in physical size and shape, it will not be possible to develop a generic template to acquire a specific category of targets (e.g., command-and-control centers) from all approach azimuths. Because of the high cost of this weapon class and the high value of the types of targets it will be employed against, very high probabilities of success will usually be required.

A different approach is being taken to detecting and identifying tactical targets, such as armored vehicles and TELs, that are stopped or in hiding. High-resolution imaging sensors[2] and ATR algorithms are being developed that will be insensitive to viewing angles. That insensitivity will be important because, in contrast to the case of fixed targets, it will be difficult to predict what the target's orientation will be when the sensor platform or weapon reaches the target area. In many cases, the range to the targets may be uncertain as well, so that the algorithms must be range-insensitive, relying more on shape correlation than size correlation.

Because the size and shape of the targets will be known in advance, a series of generic templates for a range of viewing angles could be developed. The algorithm, relying on newer and faster processors, would then cycle through the various templates to rapidly identify the objects in the image. Alternatively, it may be possible to develop a single "aspect-insensitive" template for each target type based on higher-order "moments" of each target's shape. In either case, the algorithm must be robust enough to identify targets even when they are partially obscured by foliage, other structures, or terrain features. Such algorithms are under development and have met with limited success.

The sensor and algorithm could be hosted on an aircraft or a UAV (or a satellite if the sensor produces high-resolution imagery) to aid in detection of targets of known size and shape. Presumably, the algorithm would be able to search large amounts of imagery much faster

---

[2]A high-resolution *imaging* sensor would be used because it would have a higher probability of discriminating between similar targets (e.g., fuel trucks and TELs) than non-imaging sensors (unless the target possesses some unique signature that can be exploited).

than an imagery analyst.[3] As targets are identified (e.g., as aircraft are found in revetments), their geographic positions could be communicated to weapon-delivery platforms, either in-flight air vehicles or long-range artillery. If actual imagery is communicated to the shooter with the targets annotated, this reference imagery could be used by the weapon to correlate with an onboard sensor's image of the target area. This is a form of "scene-matching," but in contrast to the downward-looking DSMAC sensor used by TLAM, the image captured by the weapon's forward-looking sensor would contain the targets. A possible operational example would be a reference image taken by the U2's synthetic aperture radar (SAR) being passed to an in-flight cruise missile that (1) images the target area with its SAR, (2) correlates with the reference image, thereby locating and identifying the targets, and (3) attacks the highest-priority targets with submunitions.

Alternatively, such a sensor and algorithm could be hosted on a smart weapon, which then directs its submunitions to their individual targets, or the sensor and algorithm could be a component of a smart submunition that is released by a "dumb" weapon. In both cases, the weapon must know that there is a high likelihood that targets are in the area; for example, it could be cued by forward observers or by other sensors (e.g., SIGINT may indicate that targets of a certain type are in a specific area).

Because of the types of tactical targets attacked, the sensor and ATR algorithm may have a lower probability of success than that for the high-value fixed targets discussed above, but the purpose is still to identify a specific target type and attack it, thereby maximizing the weapon's effectiveness.

IMPROVED GROUND VEHICLE STEALTH

Often, discussion of the application of stealth technologies and employment concepts is focused solely on aerospace systems.

---

[3]These algorithms must be robust against false target identification because they are being designed to perform a broad-area search for targets; in areas where there are no targets, they should not find any. The algorithms designed for high-value fixed targets have it much easier in this respect, since they know the target (only one) is somewhere in the image.

However, stealth is also applicable to ground-based systems. Stealth technologies may be applied to ground vehicles to reduce probability of detection by IR, radar, acoustic, and visual means. These applications of stealth technology, coupled with inputs from offboard sensors and other intelligence, may substantially enhance the abilities of stealth ground vehicles to engage enemy forces throughout the depth of the battlefield.

One concept might include the use of stealthy hunter vehicles, assisted by extended, unmanned sensors or sensor nets, to detect, classify, and locate targets and provide information to ground-based weapon batteries with smart, precision-guided standoff weapons. Using the targeting information provided by the forward-deployed stealthy hunter vehicles, the killer batteries could effectively strike ground forces well forward of the friendly main force elements.

As a first step, this or similar concepts should be examined to determine their warfighting merit. If these concepts are found to have substantial merit and a decision is made to develop them, several technology challenges will have to met. Efforts will be made to determine the signature requirements of hunter vehicles and how best to obtain them. Equally important to the success of such concepts is the development of relatively cheap offboard sensors that can be seeded to create a network to help onboard sensors to detect, classify, and locate the targets. Moreover, inexpensive low-probability-intercept communication capabilities will be required to minimize the likelihood of disclosing the position of the hunter vehicles to the adversary that has modest direction-finding capabilities.

Concepts such as these have already been examined and continue to be assessed as potentially viable means to increase force effectiveness in contingency operations. Deep attack represents a particularly promising application.

EXPANDING THE CONTRIBUTION OF STEALTH AIRCRAFT

For stealthy aircraft, there is a tension between the desire to operate during daylight hours, when targets (e.g., enemy ground forces in mass formation) may be at their most vulnerable, and the desire to operate at night, when the aircraft may be most survivable. This issue is not trivial: Concern about survivability of the stealthy aircraft can create a perverse outcome if stealthy (and presumably more-survivable) aircraft are confined to the relative safety of night operations, when they might be less effective, while less-survivable aircraft are sent to operate in the daylight.

Several broad justifications have been offered for restricting daylight activity by stealthy aircraft:

- A stealthy asset may lack a "balanced design," in which all the observables come into play at approximately the same distance from an enemy sensor platform. If the aircraft can be visually spotted much before a radar or IR sensor might detect it, avoidance of daylight operations might be imperative.

- Stealthy assets are so few in number that they must be protected for higher-value missions for which they are uniquely suited. Loss of a stealthy aircraft in a more risky daylight operation could produce a lasting effect on larger and more important operations. A variation of this argument is that a stealthy aircraft represents such a large capital investment that the additional risk of daylight operations cannot be taken. Presumably, this reflects a concern about the impact of replacement costs.

- Many of the missions that might be conducted during daylight operations might not require the particular operational and technical characteristics of the extant stealthy designs. Consequently, the exposure to the additional risk of attrition would not be deemed appropriate.

The process of deciding if a stealthy aircraft should operate during daylight hours is thus complex and highly dependent on the context of the operation. Ultimately, the judgment of the joint-force commander will be paramount. If, in the commander's judgment, a target must be attacked during the daylight, it will be attacked.

Critical questions are whether or not force packages are necessary for daylight operations, and how small those force packages can be and still keep attrition acceptable. To avoid the need for invention in the midst of war, preparations need to be made early. A good first step in the process would be to determine in a variety of contexts what types of support packages are necessary for different operations. In this process, one would want to start from having no support packages, and then add supporting forces to get the desired level of attrition. Such work requires both live testing to properly assess pilot performance in visual operations and simulation work to help screen a larger number of concepts.

The one thing to keep in mind is that attrition will be experienced in operations over hostile territory. Even low-signature aircraft are subject to bad luck in chance engagements, and enemy countermeasures will have their effect over the longer run. If stealth aircraft are too expensive to be allowed to undergo risk of attrition, the overall utility of those systems is limited. This outcome can be avoided by developing cost-saving measures based on the procurement of larger numbers of platforms, or by focusing on very different approaches that might use other PCS systems to perform the function of manned aircraft in highly contested airspace. Given the recent history of both unsteady funding and very limited production runs, perhaps it is time to begin considering those other approaches.

UNCONVENTIONAL "PRECISION STRIKE"

If *precision strike* may be viewed broadly as the ability to disable targets behind enemy lines with high confidence, physical weapon delivery may not always be necessary. We are referring here to the case in which the targets are information system elements.[4] In viewing the opportunities and risks of information warfare (IW) as an element of U.S. conventional strike, it is important to understand that IW's applicability is related to the sophistication of a potential adversary and is conditioned by the relative disparity between U.S. and enemy forces. Nations not having the necessary infrastructure to exploit information systems are not easily affected by information warfare. Very sophisticated nations build countermeasures into their systems that protect their assets from attack, and they include a great deal of redundancy, which makes successful attack difficult. Nations falling in between are the ones that are vulnerable to information warfare activities. This fact has to be taken into account when considering these approaches to warfare.

---

[4]These could be elements of military systems or of civil systems. Attacking the latter may not achieve the desired goal and may have unpredictable side effects, so we restrict ourselves here to the former.

There are many approaches to "attacking" information elements. The simplest is disruption of data pathways to limit the timely flow of data. It can be accomplished physically through cutting of pathways or jamming, through disruption of sensory apparatus, or through interference with the virtual networks underlying systems. A second approach might include attacks against the data itself through a process of corruption. A third approach might include attacks against the software manipulating the data or controlling systems. And finally, a high-level approach would treat the entire C3I/ decisionmaking process as a whole and attempt to perturb its basic functions by creating exploitable delays in systems as well as by misdirecting enemy activities.

Information warfare is not likely to offer by itself a decisive advantage across all levels of conflict. It can help other operations, but information warfare attacks in themselves may not have the impact necessary to alter enemy actions. Too few U.S. adversaries are likely to be of the appropriate level of sophistication to lend themselves to attack, and protective countermeasures, once developed, will be easily deployed. Software-based protection mechanisms, unlike those to protect hardware, can be rapidly propagated.

This brings up a final point: The United States needs to develop an aggressive defensive program for both military and civil systems. Any realistic program will be challenging, since it must consider the development of more-secure operating systems, strong encryption and authentication mechanisms, and the need to preserve usability and user acceptance. (An example is the recent Clipper chip debate.) Such a program would effectively begin to protect computer and telecommunications systems from attack and eavesdropping at a time when the United States itself may want to engage in such activities. This area will require a great deal of additional analysis before a proper balancing of interests can be addressed.

BIBLIOGRAPHY

Chu, David, "Refocusing the 'Roles and Missions' Debate," November 1994 Schulze Memorial Essay, *Marine Corps Gazette.*

Commission on Roles and Missions of the Armed Forces, *Directions for Defense,* Washington, D.C., May 24, 1995.

Hura, Myron, and Gary McLeod, *Intelligence Support and Mission Planning for Autonomous Precision-Guided Weapons: Implications for Intelligence Support Plan Development,* MR-230-AF, RAND, Santa Monica, Calif., 1993.

Pace, Scott, et al., *The Global Positioning System: Assessing National Policies,* MR-614-OSTP, RAND, Santa Monica, Calif., 1995.

Thaler, D. E., *Strategies-to-Tasks: A Framework Linking Means and Ends,* MR-300-AF, RAND, Santa Monica, Calif., 1993.

U.S. Joint Chiefs of Staff, *Functions of the Armed Forces and the Joint Chiefs of Staff,* April 21, 1948.

Watman, Kenneth, et al., *U.S. Regional Deterrence Strategies,* MR-490-A/AF, RAND, Santa Monica, Calif., 1995.